ASTROMETRIC BINARIES

F. W. BESSEL

(1839)

Astrometric Binaries

An International Conference to Commemorate the Birth of
Friedrich Wilhelm Bessel (1784-1846)

Edited by

ZDENĚK KOPAL
University of Manchester

and

JÜRGEN RAHE
Remeis-Observatory, Bamberg

Reprinted from
Astrophysics & Space Science, Vol. 110, No. 1

D. Reidel Publishing Company
Dordrecht / Boston

ISBN-13: 978-94-010-8858-9   e-ISBN-13: 978-94-009-5343-7
DOI: 10.1007/978-94-009-5343-7

TABLE OF CONTENTS

Astrometric Binaries

PART III: ASTROMETRIC BINARIES – ASTROPHYSICS

PART IV: HELIOMETRIC ASTROMETRY

PREFACE

During the past years, a number of international astronomical conferences were held at the Remeis-Observatory in Bamberg, four of them sponsored by the International Astronomical Union. The first meeting was organized in 1959 and dealt with Variable Stars, the last one was held in 1981 and focussed on 'Binary and Multiple Stars as Tracers of Stellar Evolution'.

The present conference was organized to commemorate the 200th anniversary of the birth of Friedrich Wilhelm Bessel, who was born in Minden on July 22, 1784, and died in Königsberg on March 17, 1846.

When the plan for an international conference on astrometric binaries was presented to several colleagues, we received enthusiastic support and decided to pursue the idea. A Scientific Organizing Committee was soon established, consisting of:

Z. Kopal	*Manchester, U.K.* (Chairman)	S. M. Gong	*Nanjing, China*
		M. Grewing	*Tübingen, F.R.G.*
V. Abalakin	*Pulkovo, U.S.S.R.*	P. v. d. Kamp	*Amsterdam, Netherlands*
J. Dommanget	*Uccle, Belgium*	M. Kitamura	*Tokyo, Japan*
M. G. Fracastoro	*Torino, Italy*	J. Rahe	*Bamberg, F.R.G.*
W. Fricke	*Heidelberg, F.R.G.*	Ya. Yatskiv	*Kiev, U.S.S.R.*
E. H. Geyer	*Bonn, F.R.G.*		

The meeting took place in Bamberg at the Remeis-Observatory, Astronomical Institute of the University Erlangen-Nürnberg, from June 13 to 15, 1984.

The following institutions generously supported the meeting: Deutsche Forschungsgemeinschaft, Bonn; Stadt Bamberg; Universität Bamberg; Universität Erlangen-Nürnberg; University of Manchester.

ZDENĚK KOPAL
JÜRGEN RAHE

FRIEDRICH WILHELM BESSEL – AN APPRECIATION*

Department of Astronomy, University of Manchester, England

(Received 21 August, 1984)

It has been a part of the tradition of our profession to commemorate on suitable occasions great men of the past of our science – not only to give public expression of our gratitude for what they have done for our science, but also to draw inspiration from their accomplishments to carry on further the work which they initiated, or nobly advanced. And the present conference will belong to this tradition; for we are met in Bamberg today to commemerate the 200th anniversary of the birth of Friedrich Wilhelm Bessel – one of the greatest astronomers of new age; and one who, together with Johann Müller (Regiomontanus; 1436–1476), Johannes Kepler (1571–1630), Friedrich Wilhelm Herschel (1736–1822) and Karl Schwarzschild (1873–1916) belongs to the illustrious group of path-breakers of our science that Germany gave to the world in the past five hundred years.

Two of these carriers of sacred fire – though born in Germany (Kepler on 27 December, 1571 in Weil der Stadt; and Herschel on 15 November, 1738 in Hannover) – were destined to accomplish their life's work outside the ethnic frontiers of their native land; and while Kepler returned there – in extremis – to die in Regensburg on 15 November, 1630, only Herschel (28 August, 1822) found his final resting place – near Windsor in England – far from the land of his birth (though dynastically connected with it at that time) – which became also his home for most part of his long life.

But the parallel between the two great men goes further. Both Kepler – a great theoretician, and Herschel – an equally great observer – were prone to 'jump to conclusions' which sometimes (though not always!) took them far in advance of their age. In contrast, Bessel as well as Schwarzschild combined in one person the gifts of outstanding observers with those of sublime theoreticians; and while Bessel became the founder of modern astrometry, almost the same could be said of Schwarzschild in the field of astrophysics.

But to return to Bessel – to whose memory we wish in particular to pay our reverence today – he was born on 22 July, 1974 in Minden, Westfalen; and apart from a few trips abroad (the last one of which was to Manchester in England) he spent his entire life in Germany. The seat of his most important scientific triumphs was the observatory in Königsberg in East Prussia, where he also died in 1846 and is buried (as Geheimrat

---

* Opening address delivered on 13 June at the International Conference on Astrometric Binaries, held at the Remeis Sternwarte in Bamberg between 13–15 June, 1984, to commemorate the 200th anniversary of the birth of Friedrich Wilhelm Bessel (1784–1846).

Ritter Bessel) not far from the site of his observatory. This was the place from which the parallax of the fixed stars was first measured with a significant result. As such (and for other reasons as well) Königsberg deserves indeed an especial place in the history of our science; and a few words on its own history may not be out of place on this occasion.

The origins of Königsberg go back to the second half of the 13th century – the time of a great eastward migration of the German people – and was founded by the Bohemian King Přemysl Otakar II as a Christian outpost on the shores of the Baltic sea (hence, its name); and for more than 100 subsequent years it remained as such under the control of the Order of Teutonic Knights. After the battle of Grunwald in 1410 the whole region came under the suzerainty of the Polish kings; but reverted later to the hands of the Electors of Brandenburg (later to become the Kings of Prussia). Since 1871 it became a part of the second German Empire; between 1918–1933, of the German republic; and between 1933–1945, of the infamous 'Third Reich' which lost it in 1945 to the Soviet armies; and thus – unknown to Bessel – Königsberg became the present Kaliningrad.

Bessel's contributions – not only to astronomy, but also to many other branches of modern science – covered so many fields as to make it impossible to survey them in a three-day colloquium*; and time limitation alone will prevent us from paying hommage to some aspects of his work by which Bessel's name is best known to the world – even among those who have no idea that Bessel was an astronomer; and that all his work was inspired by astronomical problems. I refer, in particular, to the Bessel functions – a by-word in many branches of physics and electrical engineering for the past hundred years – which had their roots in the theory of motion of two self-attracting bodies in the time-domain, and were introduced to relate the true anomaly of such bodies in their orbits with the time – a problem having nobody lesser than Johannes Kepler for its godfather!

The first particular cases of infinite series defining Bessel functions go back, to be sure, long before the time of Bessel himself: in fact, to John Bernoulli (1694) who corresponded extensively about their properties with Leibniz; and for their further discussion cf. also Daniel Bernoulli (1738). The first solution of Kepler's equation in terms of what we call Bessel's coefficients today was constructed by Lagrange in 1770; but such coefficients were encountered repeatedly by Laplace, Poisson, Fourier, and others in the course of their efforts to construct the solutions of Laplace's, Poisson's, and other equations in cylindrical coordinates. Bessel was, however, the first to formulate (in 1824) these functions in *integral* form – a feat which subsequently enabled Lommel (1868) to generalize their definition for arbitrary (non-integral) values of their indices – and one which provides a full justification for connecting such function permanently with Bessel's name. Bessel was, in fact, an applied mathematician par excellence – and more than one other type of formulae (in particular, in the theory of interpolation) carry his name up to this time.

---

* It is understood that the German Astronomische Gesellschaft plans to hold a national celebration in honour of F. W. Bessel in Minden (Bessel's birthplace) between 11–14 September, 1984 – with a wider programme – at which also other aspects of Bessel's work will be discussed.

The extent to which problems having its roots in astronomy influenced a development of special functions of applied mathematics is attested not only by the role which Bessel functions – and cylindrical harmonics in general – have played in our science ever since. Clairaut's theory of the figures of equilibrium of the Earth and the planets in the 18th century played a similar role in stimulating a development by Legendre (1793) and Laplace (1825) of a whole theory of spherical harmonics, which later found important applications also in quantum mechanics and elsewhere; while, more recently, certain

ASTRONOMISCHE NACHRICHTEN.
N⁰. 365. 366.

Bestimmung der Entfernung des 61sten Sterns des Schwans.
Von Herrn Geheimen-Rath und Ritter Bessel.

Als es Bradley gelungen war, seine Beobachtungen in Kew und Wanstead, welche die Entdeckungen der Aberration und Nutation herbeiführten, durch diese allein genügend zu erklären, ohne dazu der Annahme einer jährlichen Parallaxe der beobachteten Fixsterne zu bedürfen, ließ er nicht unbemerkt, daß ein über eine Secunde betragender Werth derselben, den Beobachtungen der Sterne γ Draconis und η Ursae majoris nicht entgangen sein würde. Indem er hinzusetzt, daß diese Sterne mehr als 400000 Mal so weit als die Sonne von uns entfernt seien *), geht hervor, daß er unter jährlicher Parallaxe den Winkel versteht, welchen die ganze Erdbahn an den Sternen einschließt.

Hierauf beruhet die später gewöhnlich gewordene Annahme, daß die jährliche Parallaxe der Fixsterne im Allgemeinen sehr klein sei. Wenn diese Annahme aber auch für die große Mehrheit der zahllosen Sterne dieser Art unbezweifelbar ist, so ist doch eben so wenig zu bezweifeln, daß einige darunter weit näher sind, als die große Menge der übrigen; bis zu welcher Grenze die jährliche Parallaxe dieser näheren Sterne steigen kann, kann aus der von Bradley erkannten Kleinheit derselben für die beiden angeführten Sterne (denen man noch mehrere andere, bei derselben Gelegenheit beobachtete hinzusetzen kann), offenbar nicht gefolgert werden. Wenn man also auch des Mittels entbehrte, durch fortgehende Verbesserung der Apparate und Beobachtungsmethoden, Größen bestimmbar zu machen, welche die von Bradley angegebene Grenze der jährlichen Parallaxen jener Sterne nicht überschreiten, so würde man dennoch die Hoffnung nicht verlieren, das Maaß der Entfernungen anderer Sterne aus den Beobachtungen hervorgehen zu sehen.

Bei dem jetzigen Zustande unserer Kenntnisse des Weltgebäudes können wir nur zwei, in der That nicht sichere Gründe der Vermuthung, daß ein Fixstern verhältnißmäßig nahe sei, anführen: nämlich den optischen Grund, seine ausgezeichnete Helligkeit, und den geometrischen, seine ausgezeichnet starke eigene Bewegung. Daß beide täuschen können,

*) Rigaud Miscellaneous works and Correspondence of James Bradley. Oxford 1832. p. 15.

16r Bd.

ist nicht zu bezweifeln; allein wenn eine Untersuchung über die jährliche Parallaxe eines Fixsterns unternommen werden soll, so sind sie dennoch die einzigen, welche seine Wahl leiten können.

Bekanntlich ist die jährliche Parallaxe einiger Sterne der ersten Größe der Gegenstand mehrerer neueren Untersuchungen gewesen. Piazzi fand im Jahr 1805 beträchtliche, von 2" bis 10" gehende Werthe dieser Parallaxen für α Tauri, α Canis maj., α Canis min. und α Lyrae, dagegen verschwindende für α Aurigae, α Bootis und α Aquilae: er selbst war mit der Sicherheit, mit welcher seine Beobachtungen diese Resultate ergaben, zwar nicht zufrieden, hielt aber einen Werth der jährlichen Parallaxe von α Canis maj. von 4" für wahrscheinlich. Sein Resultat für α Lyrae (2") wurde von dem von Calandrelli, aus Zenithsector-Beobachtungen in Rom gezogenen (4"4) noch übertroffen. Obgleich diesen Bemühungen zur Kenntniß der jährlichen Parallaxen einiger Fixsterne zu gelangen, genügende Sicherheit nicht beigelegt werden kann, indem Piazzi die seinigen selbst verdächtig macht, und das von Calandrelli angewandte Instrument nicht geeignet ist, großes Zutrauen zu seinen Leistungen zu erwecken, so standen sie doch ohne Widerspruch, und man konnte wirklich den Beobachtungen, welche zu ihnen geführt hatten, nichts außer ihnen selbst liegendes entgegensetzen. Indessen hatten die Beobachtungen der Unterschiede der Geradaufsteigungen der Sterne, seit Bradley, nicht nur eine große Vollkommenheit erreicht, sondern es war auch eine so große Zahl von ihnen, durch Bradley und Maskelyne bekannt geworden, daß man darauf eine Untersuchung gründen konnte, deren Resultat wenigstens so viele Sicherheit versprach, daß sich auch beträchtlich kleinere jährliche Parallaxen, als die neuerlich angegebenen, dadurch bestätigt oder widerlegt finden mußten. Ich suchte daher alle von Bradley, in dem Laufe von 12 Jahren, auf der Greenwicher Sternwarte beobachteten Geradaufsteigungsunterschiede von α Canis maj. und α Lyrae auf, indem sich, wegen ihrer Annäherung an 180°, in ihnen die Summe der Parallaxen beider Sterne verrathen mußte; es fanden sich 207 Beobachtungen dieser Art und sie ergaben die Summe der Parallaxe von α Canis maj. und der mit 1,227 multiplicirten von α Lyrae = 0"044 und den wahrscheinlichen

5

Fig. 1. The front page of the double issue of *Astronomische Nachrichten* (No. 365–366; dated 1838), in which Bessel reported the first reliable determination of the parallax of a fixed star (61 Cygni).

problems arising in celestial mechanics led Henri Poincaré to lay the foundations of modern topology!

But for us – congregated on this occasion – Friedrich Wilhelm Bessel had primarily been an astronomer; and that particular part of his work of greatest interest to us has been in the domain of *double stars*; a discussion of which will constitute the principal part of this conference. Others may wonder why we did not focus more attention on what is perhaps Bessel's best known achievement – namely, a first significant determination of the parallax of an (albeit double, and maybe multiple) star 61 Cygni (see Figure 1). A quest for the parallax of the stars was on, to be sure, all the way from the days of Tycho Brahe – 200 years before Bessel himself was born. Bessel was, moreover, not the only 'parallax hunter' of his age; for almost simultaneously with him, Friedrich Georg Wilhelm Struve (1793–1864) at Dorpat Observatory set his sights for this purpose on Vega (α Lyrae); and – unknown to them both – Thomas Henderson (1798–1844) at Cape of Good Hope on α Centauri. Bessel's choice was a relatively faint star (of 5.6 apparent mag.), the proximity of which was indicated by an unusually large apparent proper motion ($5''.21$ per annum); while Vega or α Centauri attracted attention because of outstanding apparent brightness.

As it happened, Henderson commenced his observations with a mural quadrant in 1981; Struve (using a 9-inch refractor equipped with a filar micrometer) since 1835; while Bessel utilized in 1838 a $6\frac{3}{4}$-inch heliometer – a type of instrument of his own design, which remained unrivalled in precision for precise measurement of large angles until the advent of photography. The first results of this work to appear in print were those by Struve, who in his *Mensurae Micrometricae* (1837) reported the annual parallax π of Vega to be $0''.125 \pm 0.055$ (p.e.) – a result known to Bessel when he commenced his observations of 61 Cygni in 1838; but before the end of that year he was able to announce (cf. Bessel, 1838; see Figure 1) the parallax of this star to be equal to $0''.314 \pm 0''.014$ (p.e.). As for Henderson, although he was the first one to commence the observations of α Centauri which led to a determination of its distance, he was the last to publish (in 1840) his result of $1''.16$ for its parallax*.

The important fact was, to be sure, not so much which parallax was determined or published first, but which parallax actually dispelled all doubts of contemporary astronomers that the long-sought-for annual motion has finally been significantly established. And there can be no doubt that it was the parallax of 61 Cygni, and not that of α Lyrae or α Centauri, which actually provided this assurance; for it was affected by a probable error of only $\pm 0''.014$ to be applied to a parallax of $0''.314$ as established by Bessel (its best modern value being $0''.299 \pm 0''.003$); whereas the parallax of Vega is now known

---

* We have it on the authority of the late Professor W. M. Smart of Glasgow (cf. Smart, 1950) that Henderson was in no hurry to reduce his observations because he was near his retirement age, and intended to carry out the reductions at his leisure after his return to Scotland; it was only a previous appearance of the work by Struve and Bessel which prompted Henderson to speedier action. If so, his somewhat dilatory tactics was an accurate forerunner of what happened with the discovery of Neptune a few years later!

to be 0″.123; and that of α Centauri, 0″.760 – its distance being one-third larger than that assigned to it by Henderson.

The deep impression which Bessel's memorable achievement created in astronomical circles was reflected in the following piece of sonorous Victorian prose, taken from an address by Sir John Herschel on the occasion of the award of a Gold Medal by the Royal Astronomical Society of London to Bessel in 1841:

"I congratulate you, and myself, that we have lived to see the great and hitherto impassable barrier to our excursions into the sidereal Universe – that barrier against which we chafed so long and so vainly – almost simultaneously overleaped at three different points. It is the greatest and most glorious triumph which practical astronomy has ever witnessed ... such results are among the fairest flowers of human civilization."

And what would Sir John – or, for that matter, Bessel himself – have said that they known that the apparent motion of 61 Cygni in the sky is influenced by more than the annual parallax; and that the brighter component of 61 Cygni seems to be exhibiting also periodic deviations from rectilinear motion in a period of 4.6 yr – and caused (possibly) by the presence in its neighbourhood of the body of planetary – rather than stellar – order of magnitude (cf. Van de Kamp, 1973)!

A discovery of unseen companions of nearby stars, whose presence is disclosed by periodic fluctuations of their position in the sky – rather than by the light emitted by them – represents another one of Bessel's signal contributions to astronomy; and its story goes back to the beginning of the 18th century, when Edmond Halley (1718) noted that a few bright stars must have changed their apparent positions in the sky since the time of Hipparchos by more than the angular diameter of the Sun or the Moon; and one of them happened to be Sirius (α CMa) – the brightest star of the heavens. Subsequent work only confirmed Halley's conclusions; and the proper motion of Sirius was found to amount to 1″.32 per annum (i.e., 40 min. of arc in 1800 yr); and a slightly slower proper motion (1″.25 per annum) was subsequently established for Procyon (α CMi).

The positions of both these stars were kept under continuous observation throughout the 18th century; and it soon transpired that their apparent motions in the sky were not rectilinear. The 18th century astronomers were content to record this fact; but it was left to Bessel's genius to fathom its cause: namely, to establish that both Sirius and Procyon are double stars – consisting of a visible star attended by an invisible companion – and that the periodic fluctuations in positon of the former are caused by its absolute motion around the system's centre of mass.

Bessel seems to have arrived at this conclusion in 1834 (taking as evidence the observations in right-ascension only; as these appeared to be more reliable to him than those of the declination); and by 1844 developed a complete theory of this phenomenon. 'I adhere' – wrote Bessel by that time in his famous letter to Alexander von Humboldt (cf. pp. 237ff of Aitken, 1936) "– to the conviction that Procyon as well as Sirius are genuine binary systems, each consisting of a visible and an invisible star... We have no reason to suppose that luminosity is a necessary property of cosmic bodies. The visibility of countless stars is no argument against the invisibility of countless others."

The subsequent history of Sirius and Procyon amply confirmed Bessel's prescience. Bessel – and, following him, Peters (1851) – utilized the fluctuations observed in the right-ascension of Sirius, and in the declination of Procyon, from which only elements of the relative orbits of these stars could be established. A determination of their absolute orbits – including the individual masses of their components – could not be undertaken until the parallax of these systems have been established. This was apparently done first by Elkin and Gill in the second half of the 19th century (its modern values being $0\rlap{.}{''}375$ for Sirius, or $0\rlap{.}{''}288$ for Procyon); and the absolute masses given by Auwers: their latest values (cf. Van de Kamp, 1971) being $m_A + m_B = 2.3 + 0.98 \odot$ for Sirius and $1.74 + 0.63 \odot$ for Procyon.

By the time when Auwers completed his work, Sirius B of course ceased to be an 'invisible' star; having been visually spotted long before (1862) by Alvan Clark in Cambridge, Mass. in the course of optical tests of the new 46-cm objective made for Dearborn Observatory in Evanston, Illinois*; while the visual detection of the companion to Procyon (by Schaeberle in 1896) – much fainter than that of Sirius – had to await a completion of the 91-cm refractor of Lick Observatory at Mt. Hamilton in California.

But – all considered – the strange story of the companions of Sirius as well as Procyon throughout the 19th century remained an almost European (in fact, German) affair; though their eventual visual detection was accomplished in the United States (there were no telescopes of sufficient optical power in Germany available to do so at that time). On the whole, the picture of these two systems we inherited from the 19th century was nothing much out of the ordinary; and foreshadowed little of what was in store for us when large reflectors of the 20th century were turned on those two of our close celestial neighbours. A considerable difference between the apparent brightness of the two components of these systems did not seem at first unduly surprising; for with their primary components being typical Main Sequence stars, were not their companions typical red dwarfs?

However, in 1914 – shortly before the European continent was engulged in the first of its Great Wars, which sent most European astronomers to other preoccupations (and Karl Schwarzschild to his death on the Eastern Front) – Walter S. Adams (1915) turned on the companion of Sirius a spectrograph attached to the 60-inch reflector of Mt. Wilson Observatory – to find that this companion was no red dwarf at all, but a star of the A-type, of surface temperature at least twice as high as that expected for a red dwarf. But in order to reconcile this temperature with the observed apparent brightness of the respective object, it was necessary to diminish its dimensions to a planetary size, and compress its material to densities unheard of before – the first 'white dwarf' was

---

* A large part of the reason why Bessel's 1845 prediction of the existence of the companion of Sirius took 18 yr for it to be verified by the observations was due to particular features of its orbit of a period of 49.94 yr. Already Bessel was aware of the fact that the orbit of Sirius must be highly eccentric (its modern value being $e = 0.588$); and the semi-major axis of its relative orbit $a = 7\rlap{.}{''}62$. As a result, the apparent separation of both components can fluctuate between $3\rlap{.}{''}14$ and $12\rlap{.}{''}10$ every half-century. Since, moreover, the longitude of periastron $\omega = 145\rlap{.}{°}9$, by the time of Bessel's first prediction this separation was close to its minimum; and not far from its maximum when Alvan Clark spotted the companion in 1862.

born in front of our eyes as the invisible star whose existence was predicted by Bessel in 1834.

And it was not to remain the only object of this kind for long. In the course of the time which elapsed since 1914, many hundreds of white dwarfs were detected in binary systems – close (V471 Tauri; Nelson and Young, 1970) as well as wide (Luyten, 1970; or Agayev *et al.*, 1982); and (with observational selection duly taken into account) it rapidly transpired that white dwarfs à la companions to Sirius or Procyon – far from being exceptional – must represent a typical stage of stellar evolution. The implications of such a situation are far-reaching (cf., e.g., Kopal, 1984); but their discussion must remain wholly outside the scope of this address – except to reiterate what a priceless heritage we received in them from Friedrich Wilhelm Bessel's discovery – by logic, if not by eyesight – of invisible companions in double-star systems more than a century before our time; and of ever-increasing interest.

May I only be permitted to make one remark, in this connection, suggesting that the story of the companion may still be far from closed at this time. As it has become clear since the work of Chandrasekhar (1935) a conversion of a post-Main-Sequence star into a white dwarf must be connected with a large loss of mass of the respective star; and the question arises: what has happened to this mass – and where is it now? A recent discovery of an X-ray sources in the proximity of Sirius (cf. Chlebowski *et al.*, 1981); or of a peculiar flare star near Sirius (cf. Sharma *et al.*, 1983) should perhaps lead us once more to reconsider this question. Both these objects – separated from Sirius A by 6–9 arc min (but, incidentally, by much less from each other) – cannot be any nearer to it than 1000 or more astronomical units; and according to the opinion of the astronomers just quoted they are very probably background objects separated from Sirius in space by many parsecs.

This may well be the case; but if so, the whereabouts of the 'missing mass' continues to confront us with an open problem. As the minimum distance between the new sources and the Sirius AB system cannot be less than 1000 AU, the orbital period of such a hypothetical third body would be much too long to disclose any indications of orbital motion still for many centuries to come. The only practicable way to establish their possible physical association with Sirius would be by a commonality of their apparent proper motions. It is to be hoped that some attention to these may be paid in the near future; for should they – against all expectations – prove to be the same, the whole problem of the mass-loss accompanying the formation of a white dwarf could still take an unexpected turn.

This – and much else – will no doubt be discussed in forthcoming sessions of our meetings in the next three days. Before we do so, however, it should be of interest to us all – and eminently appropriate as well – to learn more about the personality and life of the great astronomer whose accomplishments should fill us all with pride – especially so as we have an eminent colleague with us to guide us on this pilgrimage: namely, Professor Dr Walter Fricke of Heidelberg University, Director of the Astronomisches Recheninstitut of Heidelberg, and the senior astronomer of Germany who devoted his life to further development of those parts of astronomy which are almost synonymous with Bessel's name: Professor Fricke.

References

Adams, W. S.: 1915, *Publ. Astron. Soc. Pacific* **27**, 236.

Agayev, A. G., Guseinov, O. H., and Novruzova, H. I.: 1982, *Astrophys. Space Sci.* **81**, 5.

Aitken, R. G.: 1936, *The Binary Stars*, McGraw Hill Co., New York.

Bernoulli, D.: 1738, *Comm. Acad. Sci. Impér. Petropolis* **6**, 108.

Bernoulli, J.: 1694, *Acta Eruditorum Lipsiae*, p. 435.

Bessel, F. W.: 1824, *Untersuchung des Theils der Planetarischen Störungen, welcher aus der Bewegung der Sonne entsteht*, Berlin Abhandlungen.

Bessel, F. W.: 1838, *Astron. Nachr.* Nos. 365–366; see Fig. 1.

Bessel, F. W.: 1845, *Astron. Nachr.* **22**, 145, 169, 185.

Chandrasekhar, S.: 1935, *Monthly Notices Roy. Astron. Soc.* **95**, 207, 226, 676.

Chlebowski, T., Halpern, J. P., and Steiner, J. E.: 1981, *Astrophys. J.* **247**, L35.

Halley, E.: 1718, *Phil. Trans. Roy. Soc. London* **29**, 454.

Henderson, T.: 1840, *Mem. Roy. Astron. Soc.* **4**, 92.

Herschel, F. J. J.: 1841, *Monthly Notices Roy. Astron. Soc.* **5**, 89.

Kopal, Z.: 1984, *Astrophys. Space Sci.* **99**, 3.

Lagrange, J. L.: 1770, *Hist. Acad. Roy. Sci. Berlin* **25**, 204.

Laplace, P. S.: 1825, *Mécanique Celeste*, Vol. 5, Paris.

Legendre, A. M.: 1793, *Mémoires de Mathématique pour 1789*, Paris.

Lommel, E. C. J.: 1868, *Studien über die Besselsche Functionen*, Leipzig.

Luyten, W. J.: 1970, *Astron. Publ. Univ. Obs. Minnesota*.

Nelson, B. and Young, A.: 1970, *Publ. Astron. Soc. Pacific* **82**, 699.

Peters, C. A. F.: 1851, *Astron. Nachr.* **32**, 1.

Sharma, D. P., Marar, T. M. K., Seetha, S., Shylaja, K. S., Kasturirangan, K., Rao, U. R., Bhattacharyya, J. C., and Rozario, M. J.: 1983, *Astrophys. Space Sci.* **91**, 467.

Smart, W. M.: 1950, *Some Famous Stars*, Longman's Green Co., London, p. 45.

Struve, F. G. W.: 1837, *Stellarum duplicium et multiplicium Mensurae Micrometricae*, Mem. de l'Acad. Imper. des Sci. de St. Petersbourg.

Van de Kamp, P.: 1971, *Ann. Rev. Astron. Astrophys.* **9**, 103.

Van de Kamp, P.: 1973, *Astron. J.* **78**, 1092.

FRIEDRICH WILHELM BESSEL*
(1784–1846)

In Honor of the 200th Anniversary of Bessel's Birth

WALTER FRICKE

Astronomisches Rechen-Institut, Heidelberg, F.R.G.

(Received 28 July, 1984)

Abstract. May I first express my warmest thanks to the organizers of the Conference for having invited me to present a brief bibliographic report on the great astronomer F. W. Bessel. You have given me a welcome opportunity to express an appreciation of Bessel's work which has marked the beginning of research on the motion of stars and which was pioneer work on astrometric binaries and on the determination of stellar distances. Bessel has established the first celestial coordinate system which approximates an inertial system, and my colleagues and I at Heidelberg in establishing an improved fundamental reference system, the FK5, have had to review again some of Bessel's pioneer contributions to this field, and such contributions shall be mentioned later. My talk shall be divided into the following parts: (1) Bessel's course of life and family; (2) Education in astronomy in Bremen; (3) Fundamenta Astronomiae; (4) Fundamental observations at the Königsberg Observatory (or according to S. Newcomb (1906): the German School of Astrometry); (5) 61 Cygni; (6) Bibliography of Bessel's original works.

1. Bessel's Life and Family

Bessel was born on 22 July, 1784 in Minden, Westphalia. His father was a civil servant of the town; his mother was the daughter of the minister Schrader of the little town Rheme in Westphalia. There were nine children, three boys and six girls. Bessel and his brothers received an education at the Gymnasium in Minden. His brothers completed their education with success and received a university education in law. Both entered positions as judges at high courts. Bessel left the Gymnasium at the age of fifteen because he had difficulty with Latin.

In 1799 Bessel entered the famous mercantile firm of Kulenkamp in Bremen as an apprentice, where he was to serve for seven years. From 1799 to 1805 he not only fulfilled excellently his duties in the firm of Kulenkamp, but also carried out studies in foreign languages, in mathematics and in astronomy. In 1806 he discontinued his apprenticeship and entered a position as an assistant astronomer at the private observatory of J. H. Schröter in Lilienthal near Bremen. In 1809 the Prussian King Friedrich Wilhelm II ordered the construction of an observatory in Königsberg, East Prussia, and Bessel was appointed its director and professor of astronomy, on recommendation of Wilhelm von Humboldt. The title of doctor, required for professorship, was awarded to him by Gauss at the University of Göttingen. The Königsberg Observatory has become Bessel's final

---

* Communication presented at the International Conference on 'Astrometric Binaries', held on 13–15 June, 1984, at the Remeis-Sternwarte Bamberg, Germany, to commemorate the 200th anniversary of the birth of Friedrich Wilhelm Bessel (1784–1846).

home. In 1812 he married Johanna Hagen. They had two sons and three daughters; but the marriage was clouded by the early death of both sons. For thirty years Bessel very seldom left his observatory. His last and longest trip, in 1842, was to England, and it was one of the most stimulating ones for him. After several years of suffering Bessel died of cancer on 17 March, 1846; he was buried near the observatory.

2. Bessel's Education in Astronomy

Self-educating himself by exclusively using textbooks, Bessel learned foreign languages and mathematics and astronomy with great success. Books on practical navigation served him as tools for carrying out determinations of time and of the longitude of Bremen. After reading the textbook by Lalande and Olbers' work on the determination of a comet's orbit from several observations, Bessel was able to carry out a reduction of observations made of Halley's Comet in 1607 and a determination of the comet's orbit. Bessel presented the results to Olbers in 1804 – in his 5th year as an apprentice in the Kulenkamp house – and Olbers in recognizing the brilliant qualifications of Bessel marked the turning point in Bessel's life. On Olbers' recommendation, Bessel's determination of the orbit of Comet Halley – on the level of a doctoral dissertation – was published; and it attracted much attention.

Within the short time of a few years at Schröter's private observatory at Lilienthal, Bessel acquired experience in observations of comets and planets, in particular, of Saturn and its rings and satellites. On Olbers's suggestion, Bessel started in Lilienthal the reduction of Bradley's observations of about 3000 stars made at Greenwich from 1750 to 1762. This work can be considered as the masterpiece and the highlight of the period of Bessel's unusual self-education in astronomy.

3. Fundamenta Astronomiae

The star catalogue containing the results of the reduction of Bradley's observations was completed by Bessel in 1818 after ten years work. The title is *Fundamenta Astronomiae pro anno 1755 deducta ex observationibus viri incomparabilis James Bradley in specula astronomiae Grenovicensi per annos 1750–1762 institutis*. As a rule, Bradley had made five observations per star in RA and four observations in DEC. The determination of refraction from these observations was probably one of the most difficult tasks. When Bessel completed in 1811 the construction of refraction tables, he was awarded the Lalande Prize of the Institut de France. The information which Bessel has included in the *Fundamenta* consists of the name of the stars, the magnitude, RA and Dec to $\frac{1}{10}$ of a second of arc for 1755, and the values of Bessel's annual precession for 1755 and 1800. The clock stars which served for the determination of the equinox and equator by means of observations of the Sun were observed about 50 times. Although Bessel was certainly the only astronomer of his time able to reduce Bradley's observations, the work must have been extremely exhausting for him. He wrote that an astronomer should never carry out more observations than he can reduce himself. In the case of Bradley's observations, however, Bessel had recognized their usefulness for providing the best possible celestial coordinate system for the epoch 1755. Second, he recognized the need of accurate first-epoch observations for the determination of proper motions of stars and for the establishment of a reference system for a large interval of epochs. In fact, in the *Fundamenta* he has already given the proper motion for each of the stars for which second-epoch observations were made around 1800 by Piazzi in Palermo, Italy. For the determination of precession from such proper motions, Bessel was awarded a prize by the Berlin Academy of Sciences. In conclusion, one may duly consider Bessel's *Fundamenta* to mark the beginning of modern astrometry.

4. Observations at the Königsberg Observatory

While the observatory in Königsberg was being built from 1810 to 1813, Bessel concentrated on the reduction of Bradley's observations. In 1813 Bessel began to observe positions of stars with two modest instruments, a Dollond transit instrument and the Cary circle, which gave an opportunity to develop with great perfection the methods of determination of instrumental errors and of atmospheric effects. Bessel as a great master of astronomical observations has had the good luck to obtain the following three instruments, build by capable instrument makers: a Reichenbach–Ertel meridian circle in 1819, a large Fraunhofer–Utzschneider heliometer in 1829, and a Repsold meridian circle in 1841.

In the observations of positions of stars, Bessel concentrated first on the application of absolute methods – there was hardly any other choice for him, because a reference system had to be established at a modern epoch. For this purpose Bessel observed Maskelyne's stars; these are 36 clock stars in the zodiacal zone which were selected and observed by Maskelyne, the successor of Bradley in Greenwich. Bessel had noticed that these bright stars which are fairly uniformly distributed in right ascension are well-suited for day-observations with the Sun, Mercury, and Venus, such that the position of the equator and the equinox can be determined.

The results of several years of observations provided a second epoch for the reference system whose positions at the first epoch were observed at Greenwich. The observations were made with the Reichenbach–Ertel transit circle, and Bessel's methods of determining instrumental errors and of eliminating atmospheric effects were applied. As the result of the observations of the 36 Maskelyne stars, Bessel published in 1830 the *Tabulae Regiomontanae* presenting the mean and apparent places of the Maskelyne stars and of the polar stars α and δ Ursae Minoris for the period 1750 to 1850.

The *Tabulae Regiomontanae* have provided the celestial coordinate system not only for one century but also beyond 1850. With the aid of this reference system all observations of the Sun, Moon, and planets made at Greenwich since 1750 could be reduced, such that these observations could be used for the theory of planetary orbits. Bessel's goal of determining the motions of the stars in such a way that their positions can be predicted for all times was, of course, only insufficiently reached.

Let us consider briefly Bessel's observations of the Sun around 1820 which were

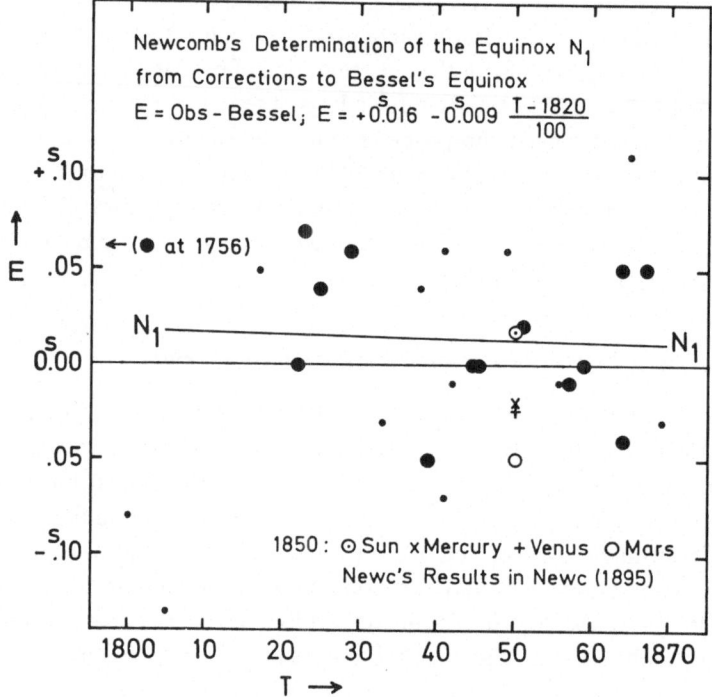

Fig. 1.   Newcomb's equinox N_1 determined from all absolute right ascensions from 1820 to 1870. The result
of Bessel's observations is adopted as reference value ($E = 0$ near $T = 1820$).

employed to determine the equinox and equator. Bessel has stated that this equinox determination should be accurate within $\pm 0\overset{s}{.}010$. When Newcomb (1872) redetermined the equinox from the results of many different observers, he formed the 'equinox corrections'

$$E = \text{Obs} - \text{Bessel},$$

which are included in Figure 1. The large filled circles represent results from observations of the Sun with highest weight and the small filled circles observations of the Sun with low weight. On the zero line at $T = 1822$ is Bessel's equinox, and the line N_1 is the result of Newcomb's least squares' determination which has shown that all determinations until 1870 combined have not yielded a significant improvement of Bessel's equinox. The deviation of the line N_1 from zero is small and of about the size of the mean error which was given by Bessel for his equinox determination. At 1850 one finds solutions for the corrections to Bessel's equinox from all observations available to Newcomb of the Sun ⊙, Mars ○, and Mercury ($\times$) and Venus ($+$), which will not be discussed here.

In Figure 2 we have compared at Heidelberg the RA of Bessel's observations of the Maskelyne stars at $T = 1825$ with the FK4 positions at this epoch. The figure shows the differences $\Delta\alpha \cos\delta$ in milliseconds as a function of declination from -30 to $+40°$

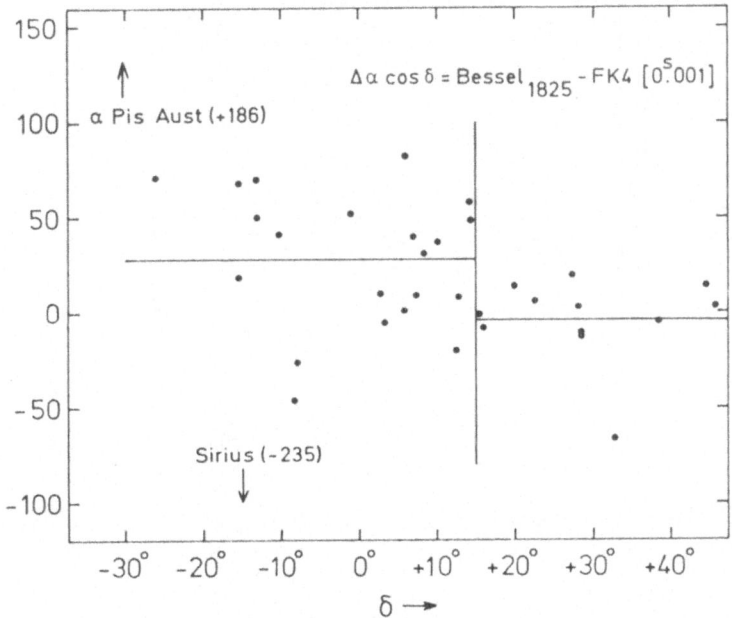

Fig. 2. Differences in RA between Bessel's observations and the FK4 at 1825.

which is the full width of the zone. To our great surprise the stars north of about $+15°$ show no systematic deviation from the FK4 and a fairly small scatter. South of $+15°$ there is a systematic deviation from FK4 of $+0^s030$ and a larger scatter. Of particular interest in the figure are the deviations in RA for Sirius (0^s235 at 1825) and α Psc Aust (0^s186 at 1825).

Bessel has remarked that he noticed a variability of Sirius' position which he suspected to be the consequence of a possible double star nature. No similar remark can be found for α Psc (Fomalhaut). We now merely known that the star is a nearby one and has a distant companion moving with equal speed but without indication of a double star orbit.

Because of the discrepancy of Bessel's observations of α Psc A in the comparison with FK4, one may hope that the phenomenon can be explained in the future.

Merely for drawing attention to the phenomenon, I quote the data given by Herczeg (1984) for α Psc A and its companion HD 216 803:

	α Psc A	HD 216 803
Parallax	$0''149 \pm 0''008$	$0''128 \pm 0''008$
μ_α	$0''386$	$0''326$
μ_δ	$-0''161$	$-0''158$
Rad. veloc.	$+6 \text{ km s}^{-1}$	$+10 \text{ km s}^{-1}$

The projected separation is about 5×10^4 AU.

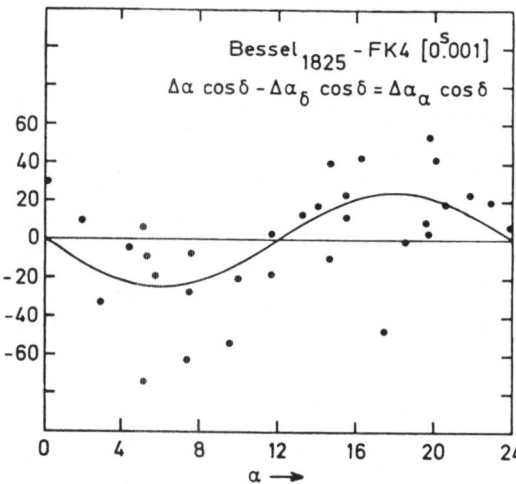

Fig. 3.   Differences between Bessel's RA and the FK4 at epoch 1825 as a function of RA.

Finally, in Figure 3 we have given the differences between Bessel's RA and the FK4 as a function of right ascensions. Here, a simple seasonal variation is clearly indicated in Bessel's RA. This is a systematic error due to the effects of the seasonal variations in the Earth's atmosphere and the clocks. It has, in fact, taken more than about hundred years, until technical means have been developed for avoiding such errors. I acknowledge the computations of the data in Figures 2 and 3 by my colleague R. Bien.

Let us now briefly review Bessel's efforts in the direction of the determination of proper motions of many stars. For this purpose Bessel carried out differential observations in zones; 'differential' means with respect to the reference system he had set up in the *Tabulae Regiomontanae* (1830).

At the Reichenbach transit circle Bessel himself observed about 75 000 stars brighter than ninth magnitude in zones from -15 to $+45°$. This program was continued by Argelander who became Bessel's assistant in Königsberg. Argelander extended this work up to $+80°$ and down to $-32°$.

From this work emerged Argelander's plan for the establishment of the *Bonner Durchmusterung*, when Argelander had become director of the observatory in Bonn. We note now that Argelander's mapping of the sky has turned out to be of utmost usefulness for more than one century.

From Bessel's zone observations emerged the plan for transit circle observations of stars brighter than 9th magnitude and distributed over the whole sky. This undertaking organized by the Astronomische Gesellschaft was carried out by about 20 observatories with mean epochs between 1875 and 1930. One must regretfully admit now that the AGK1 was an unsuccessful undertaking. The original aim of Bessel's plan of determining proper motions of many stars only recently reached a satisfactory stage in the form of the AGK2/3, which is a photographic catalogue of 180 000 stars north of $-2°$. For the southern sky, proper motions of many stars may hopefully be determined by means of astrometric satellites like HIPPARCOS.

Bessel's work on the establishment of a fundamental reference frame was extremely successful and has been a guide for further developments: certainly for Newcomb's *Fundamental Catalogue* (1898) and Newcomb's work on the improvement of the astronomical constants; furthermore, for the establishment of the series of German fundamental catalogues, the first of which was produced by Auwers (1879) in Berlin. The FK5 which will be the fifth fundamental catalogue of this series is near completion in Heidelberg.

Finally, Bessel deserves credit for the discovery of two important effects which became evident from his transit observations. Bessel noticed a variation of the declinations of stars with a quasi-period greater than one year. He explained the effect as a possible consequence of variations of the altitude of the pole. The effect was finally confirmed by Küstner in Berlin before 1890 and recognized as a consequence of latitude variations. The irregularity of these variations have made it necessary to observe the phenomenon permanently at a number of stations all over the world. As a consequence, all modern observations of positions on the sky in declination must take the latitude variations into account and, furthermore, declinations of stars measured before 1890 cannot be used anymore for the determination of stellar motions, because an extrapolation of the latitude variation into the past is impossible.

The second effect of importance noticed by Bessel in transits of the stars is the magnitude equation in right ascensions. Simply speaking, the transit of a faint star is as a rule observed too late, and the transit of a bright star is observed too early. Since the size of the effect depends on the observer, the effect is known since Bessel as the 'personal equation'. With the introduction of Repsold's impersonal micrometer and of screens of different transparency on the transit circles the effect has been minimized from about 1900 onwards.

5. 61 Cygni

The problem of measuring distances of stars had occupied Bessel for a long time. He knew that, since the time of Copernicus, learned people expected to recognize the orbital motion of the Earth in apparent circular motions of the stars. The story of the determination of the parallax of the nearby star 61 Cygni by Bessel belongs (fortunately) to the first chapters of all good introductory books on astronomy. I may, therefore, summarize here merely a few facts which are helpful for appreciating Bessel's achievement. In my opinion, there are the following facts: Bessel's successful determination was not the result of an unforeseen discovery or of good luck, but the result of extremely wise preparatory considerations and of high-quality observations by a great master. In planning the observations Bessel knew of unsuccessful attempts made already before 1800. He knew that the criteria for selecting a star suspected to be nearby were its magnitude and its proper motion, and he concluded correctly that for his attempt the size of the proper motion provides the safest criterion.

The star 61 Cygni was known to Bessel from his work on Bradley's observations which had shown two components of viz. magnitude 5.6 and 6.3. From an approximate knowledge of the orbital period of the components gained from Bradley's observations, Bessel estimated the linear distance between the components in application of Kepler's third law. The measurement of the angular distance between the components (about $16''$) and the knowledge of the linear distance allowed him to arrive at an estimate of the parallax of 61 Cygni. His estimate was $\pi = 0''.46$; according to our present knowledge the parallax is $\pi = 0''.29$. Finally, one of the most important conditions for a successful attempt of observations of 61 Cygni was the acquisition of the Fraunhofer–Utzschneider heliometer in 1829. This heliometer was in optical and mechanical outfit the most powerful instrument for the precise measurement of angular distances between stars. The angular distances of the components of 61 Cygni from two comparison stars of magnitude 9 to 10 were easily measurable with this heliometer. The distances were roughly 8 and 12 min of arc, and the measurements were extended over a period of 18 months. Bessel's result which was published in 1838 in the *Astronomische Nachrichten* was $\pi = 0''.314 \pm 0''.020$ m.e.

Recent photographical parallax determinations have yielded the value

$$\pi = 0''.292 \pm 0''.0045 \text{ m.e.}$$

demonstrating the excellence of Bessel's work. Bessel's pioneering work was supplemented by successful parallax determination in the following years:

For α Lyrae, F. G. W. Struve, Dorpat, found

$$\pi = 0''.262 \pm 0''.037 \text{ m.e.} ,$$

and for α Centauri, Th. Henderson found at the Cape Observatory a parallax of about one second of arc. Modern determinations for α Lyrae and α Centauri are $0''.121$ and $0''.75$, respectively. The pioneering work of Bessel, Henderson, and Struve around 1840 laid the foundation for studies of the structure of our stellar system. For more details, reference is made to Strassl (1946).

This review does not include Bessel's works in geophysics and mathematics. However, in the following bibliography a complete survey of Bessel's original works is given.

6. Bibliography of Bessel's Original Works

(1) *Abhandlungen von F. W. Bessel*, Rudolf Engelmann (ed.), 3 Vols, Leipzig, 1875, in: Vol. I, 23 papers on the motion of planets, 28 papers on spherical astronomy; Vol. II, 25 papers on the theory of astronomical instruments, 29 papers on stellar astronomy, 19 papers on mathematics; Vol. III, 11 papers on geodesy, 17 papers on geophysics, 33 papers on other subjects.

(2) *Beobachtungen der Königsberger Sternwarte*, 21 Volumes (1815–1840).

(3) *Fundamenta Astronomiae, 1755 ex. observationibus James Bradley Grenovicensi per anno 1750–1762,* Königsberg 1818 (title abbr.).

(4) *Tabulae Regiomontanae reductionum astron 1750–1850 computatae,* Königsberg, 1830 (title abbr.).

References

Auwers, A.: 1879, *Fundamental Catalog,* Publ. Astron. Gesellschaft XIV, Engelmann, Leipzig.

Herczeg, T.: 1984, *Astrophys. Space Sci.* **99**, 34.

Newcomb, S.: 1872, *Wash. Obs. for 1870,* App. III.

Newcomb, S.: 1898, *Astron. Pap. Wash.* Vol. 8, Paart 2.

Strassl, H.: 1946, *Naturwissenschaften* **33**, 65.

PROJECT HIPPARCOS*

P. L. BERNACCA

Asiago Astrophysical Observatory, University of Padova, Asiago, Vicenza, Italy

(Received 1 October, 1984)

Abstract. The HIPPARCOS satellite project of the European Space Agency is described in detail. The paper focusses on the basic rationale, on the performance of the scientific instrumentation and on an explanation of how scientific objectives can be achieved. The purpose is to provide quantitative visibility of basic project features to astronomers and astrophysicists not familiar with astrometric techniques by exploiting updated technical information on project parameters not generally available in the open literature.

1. Introduction

The HIPPARCOS satellite project of the European Space Agency is the result of combined effort of Advanced Space Technology and of scientific expertise in Astronomy, Geodesy, Mathematics, Optics, Automatics, Statistics, and Informatics.

It will bring Astrometry in the realm of exciting Space Sciences and it will test the feasibility of a new type of Astronomy Space Missions. For, indeed, the project will be successful only if data will be obtained continuously for at least two years of uninterrupted and optimal instrumental performance on the basis of an observing programme fully established in advance and if all science data will be globally analyzed by large interdisciplinary teams of experts.

The following presentation is restricted to the basic rationale, to the performance of the scientific instrumentation and to an explanation of how scientific objectives can be achieved. The purpose is to provide quantitative visibility of basic project features to Astronomers and Astrophysicists not familiar with astrometric techniques by exploiting updated technical information on project parameters not generally available in the open literature. A discussion of the complex architecture expected for data analysis would require more space than available and it has been preferred not to present summary qualitative descriptions, which are understandable mainly by adepts.

2. A New Start for Global Astrometry

The main role of global astrometry is the realization of a space-time reference frame to allow accurate description of astrophysical events. Such a frame should be inertial to the best achievable accuracy and it should be materialized by a large number of objects accessible to optical observations. Essential to the establishment of an inertial frame is the invariance in space of the triad to which star positions and their variation in time

---

* Communication presented at the International Conference on 'Astrometric Binaries', held on 13–15 June, 1984, at the Remeis-Sternwarte Bamberg, Germany, to commemorate the 200th anniversary of the birth of Friedrich Wilhelm Bessel (1784–1846).

Astrophysics and Space Science **110** (1985) 21–45. 0004–640X/85.15
© 1985 *by D. Reidel Publishing Company.*

are referred. The methods for the determination of reference systems have been recently discussed by Fricke (1982). The ensemble of celestial objects for which positions and proper motions are accurately known is named *Fundamental Catalogue*.

Table I presents the current status of astronomical catalogues which are adopted as

TABLE I

Accuracy in reference systems

Parameter	Mean error at present epoch (1980)	Number of stars	Faintest m_v	Catalogue
λ, β	$0''.1$	1535	7.5	FK4
μ_λ, μ_β	$0''.12$			FK4
λ, β	$0''.03$	4500	9	FK5
μ_λ, μ_β	$0''.002$			FK5
λ, β	$0''.3$	38000	10.5	IRS
μ_λ, μ_β	$0''.005$			IRS
λ, β	$0''.4$	180000	12	AGK3
μ_λ, μ_β	$0''.01$			AGK3
λ, β	$1''$	259000	~ 13	SAO
$\omega, \lambda, \beta, \mu_\lambda, \mu_\beta$	$0''.002$	116000	13	HIPPARCOS
λ, β	$0''.03$	400000	11	TYCHO

reference, based on decades of patient transit circles observations and astrographic surveys. The primary fundamental catalogue, labelled FK4 (Fricke and Kopff, 1963), contains data for about one star per $5 \times 5°$ and it is limited to 1535 stars brighter than $m_v = 7.5$. At the mean epoch 1930 the average mean error in each coordinate was about $0''.04$ and proper motions had a mean error of $0''.002$/year (this term 'year' will be omitted in the following). Error accumulation leeds to $0''.1$ at the present epoch. The FK4 system has been extrapolated to fainter stars by meridian observations of about 38000 stars. These stars, called International Reference Stars (IRS) consist of the AGK3R on the Northern hemisphere and of the SRS (Southern Reference Stars) ensemble on the Southern hemisphere. The two set overlap in a strip 10° wide about the equator. The mean accuracy of IRS is about $0''.3$ in positions and $0''.005$ in proper motions. The density of the set is one star per square degree in the magnitude range $7.5 < m_v < 10.5$. In the Northern hemisphere the AGK3 catalogue (Dieckvoss *et al.*, 1975) contains data for some 180000 stars in the range $5 < m_v < 12$ (density is 9 stars per square degree) whose positions at the present epoch, and proper motions, have mean errors of about $0''.4$ and $0''.1$, respectively. The Smithsonian Astrophysical Observatory Catalogue (SAO) contains data for 259000 stars. It is a compilation from different sources. Individual positions have standard errors of between $0''.5$ and $1''.0$ at the present epoch.

The future status will hopefully be represented by the HIPPARCOS and TYCHO catalogues foreseen as output of the HIPPARCOS mission of the European Space Agency (ESA), which is scheduled to commence early in 1988. The HIPPARCOS Catalogue will consist of about 100 000 stars brighter than $m_v = 13$ with a density of about 2.5 stars per square degree and standard errors at the epoch of observations of $0''.002$. Thus even after ten years of accumulated errors due to proper motions, the error of each position will be only $0''.02$. The TYCHO Catalogue will provide positions of at least 400 000 stars brighter than $m_v = 11$ with an accuracy of about $0''.03$. Combination of these positions with those in the AGK2 and Cape Catalogues will give a system of proper motions of about 300 000 stars with accuracies (imprecisions) of about $0''.003$.

Up to date astrometric observations have been made from the Earth and, therefore, the invariance of the reference triad is affected by the orbital and rotational motion of the Earth (eventually from tectonic displacements of the crust) and by the Moon and planetary motions. It has been essential to model all above motions (cf. Kovalevsky, 1975) in order to disentangle the effects due to a secular rotation of the reference triad from the true angular rate of change of star directions with respect to the solar system rest frame. In addition the adoption of a model for galactic rotation and for stellar kinematics has been necessary. At present, the uncertainty in the secular rotation of the stellar reference frame is of about $1''.2$ per century (in the FK4). In the forthcoming catalogue (FK5), in preparation at the Astronomisches Rechen-Institut in Heidelberg (Fricke, 1977), the uncertainty of the absolute rotation of the proper motion system will be only $0''.15$ per century. This catalogue will contain about 4500 stars brighter than $m_v = 9$ with a density of one star per $3 \times 3°$ and it is expected that random errors will be of the order of $0''.03$ in position and $0''.002$ in proper motion components.

The best way of defining an inertial system is to assign coordinates and possibly proper motions to very distant celestial objects, like galaxies and quasars, which can be considered quasi-motionless. Indeed, even if transverse velocities of galaxies were comparable with their radial velocities (100 km s^{-1} Mpc^{-1}), their proper motions would be of the order of only 10^{-5} arcsec. For a star in the direction **r** the difference between the true proper motion and that inferred from current fundamental catalogues, i.e. $\omega \times$ **r**, would permit us to derive the rotation ω of the presently adopted reference triad. Attempts in this direction are now beeing pursued by Jet Propulsion Laboratory through observations of radio cores of distant quasars and galaxies via VLBI techniques. At present the JPL VLBI system is composed of some 100 objects and the mean uncertainty in positions is about $0''.007$ (Fanselow et al., 1984). It is expected that improvement in accuracy should reach the milliarcsecond level. HIPPARCOS will provide a very consistent optical reference frame of positions and proper motions without regional errors, that is to say with almost uniform accuracy all over the sky, but the rotation of the system will be undefined unless it will be tied-up to an inertial or quasi-inertial system. Due to the magnitude limit of the telescope aboard HIPPARCOS, none of the optical counterparts of the extragalactic radio sources (there is only one exception) of the JPL VLBI reference frame can be observed in order to perform the link. However some HIPPARCOS programme stars brighter than 13th magnitude exhibit non-thermal spatially compact radio emission (e.g., RS CVn stars) and might serve as transfer objects for the link, via VLBI techniques, of their positions and proper motions with respect to angularly close extragalactic VLBI sources (Preston *et al.*, 1983). The Hubble Space Telescope will also be used to optically measure the angular separation between HIPPARCOS stars and the optical counterparts of radio-quasars (Duncombe *et al.*, 1982). In addition, the HIPPARCOS system can be linked to the FK5 system (Roeser, 1983) and to the dynamical inertial system inferable from observation of asteroids motion (e.g., Soederhjelm and Lindegren, 1982). It is expected that the final HIPPARCOS system will be inertial to better than 0″.05 per century by the combination of the above four methods (cf. Kovalevsky, 1984).

A second role of global astrometry, essential to Astrophysics and Cosmology, is the measurement of absolute parallaxes. A large increase in quality and precision of parallax data is required to solve the problem of the calibration of absolute magnitudes of stars and extragalactic objects. This calibration still today rests on the trigonometric distance of stars within about 20 pc of the Sun, which are mainly of late spectral type. For intrinsically very bright stars, e.g., Cepheids and OB supergiants, the calibration depends, at the stake, on the Hyades Main Sequence fitting and on the assumption that Hyades, α Per, Pleiades, NGC 2264, and other clusters have the same chemical composition. Most of the data currently available on trigonometric parallaxes are contained in the General Catalogue and its supplements. These together contain data for 7000 stars brighter than $m_v = 12$. Many of these stars have very small parallaxes and the average mean error is about 0″.016. Determination of relative parallaxes are systematically obtained at the U.S. Naval Observatory in Flagstaff with an accuracy of about 0″.004 (Westerhout and Hughes, 1982). The rate of increase is some 50 stars per year and the total number of parallaxes so far obtained is about 700. Relative parallaxes are also obtained in other observatories via photographic techniques and new techniques like optical interferometry are also implemented (cf. ESA, 1979; and Kovalevsky, 1984). All programmes may be compared unfavourably with the output expected from HIPPARCOS: absolute parallaxes of some 10^5 stars in few years down to $m_v = 13$ and accuracy in the milliarcsecond range.

The possible consequences of the project on the advance of astronomy and astrophysics have been summarized in the phase-A study report by ESA (1979a) and extensively presented to the scientific community in the books edited by Barbieri and Bernacca (1979) and by Perryman and Guyenne (1982). The project will also be the next major breakthrough in astronomical techniques. Figure 1, redrawn from an article by Hughes (1984) present the evolution of the error in angular measurements all over the centuries. Major technical improvements in the past were: (i) improvement of the radial scale read-out in quadrants and sextants around 1600 A.D., (ii) introduction of the telescopic sight and of the micrometer at the end of the 17th century, and (iii) replacements of quadrants and sextants with full circles at the end of the 18th century. The next step forward requires the use of a space platform in order to avoid the limitation on accuracy induced by atmospheric turbulence and refraction and to have access to the whole sky from a single observing site.

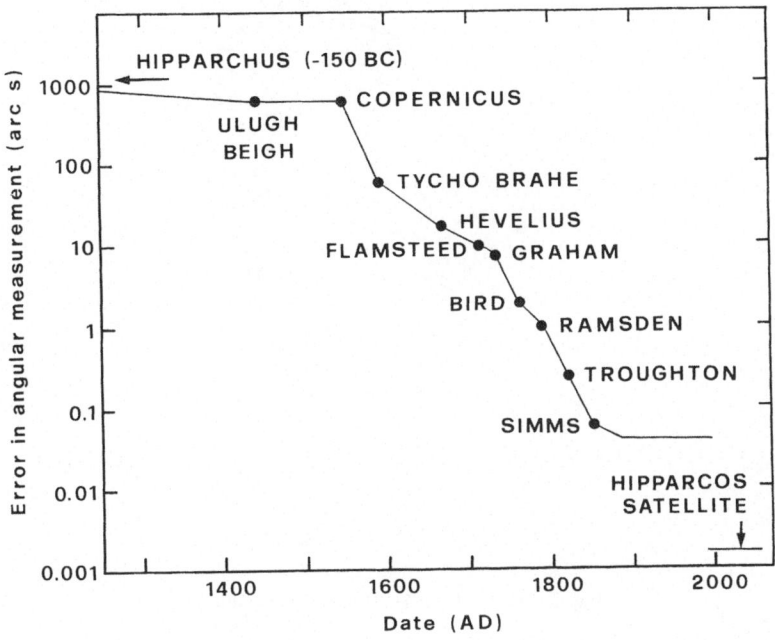

Fig. 1.   Evolution of astronomical angular accuracy.

A possible project in space astrometry was announced by Lacroute (1968) and discussed in greater detail by Bacchus and Lacroute (1974). The methods to be used for the execution of an astrometry mission and preliminary possible technical solutions were defined by the Mission Definition Group of ESA in 1975/1976 (ESA, 1977). Astrometrists in this starting team were: W. Brouw, M. G. Fracastoro, E. Høg, and P. Lacroute. In-depth analysis was made during the phase-A study in 1977/1978 which was entrused by ESA to AERITALIA for hardware and mission analysis aspects (Bernacca and Cornelio, 1978) and to AML/ACM for evaluation of the theoretical accuracy (Larcher, 1978; Boissières and Larcher, 1978). An extension of the phase-A study (ESA, 1979b) made the project ready for approval in March 1980. The final study, design and development is now contracted out by ESA to the European Consortium MESH with MATRA and AERITALIA as prime and co-prime contractors, respectively. During the final study the TYCHO programme was implemented upon proposal by Høg (Høg et al., 1982). The satellite view is shown in Figure 2. The basic principle (due to Lacroute) is to scan continously and systematically the entire sky with a telescope capable of performing differential measurements of star directions by simultaneous observations of star fields separated by a large angle.

The satellite will be launched in the first quarter of 1988 from Kourou, French Guyana, by Arianespace and placed in geostationary orbit. A nominal mission duration of 2.5 years is foreseen.

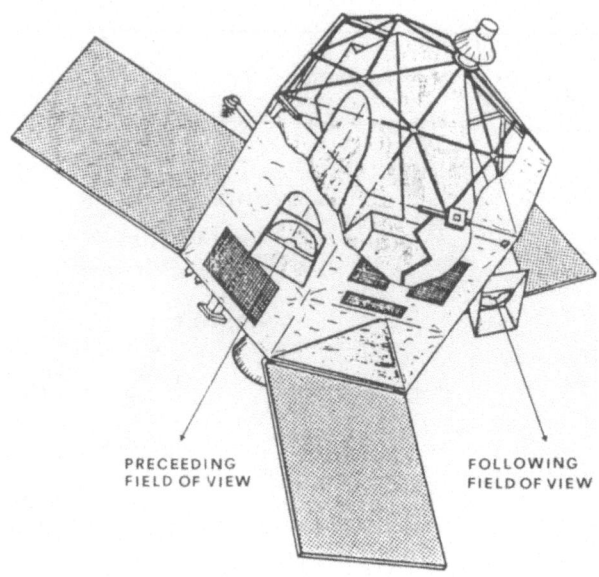

Fig. 2.   View of the HIPPARCOS satellite.

3. The Hipparcos Mission

3.1. Concepts

We are given a population of N stars on the unit sphere, each star moving in a given reference frame (e.g., the ecliptic) according to the law

$$\mathbf{r}(t) = \dot{\mathbf{r}}_0(t)t + \omega \mathbf{h}(t),\tag{1}$$

where $\mathbf{r}_0$ is the position seen from the geocenter at a given epoch, $\dot{\mathbf{r}}$ is the proper motion, t is the time, and $\omega \mathbf{h}(t)$ accounts for the parallactic displacement of a star having parallax ω. The vector $\mathbf{h}(t)$ is tangent to the sphere at $\mathbf{r}$ (in practice at $\mathbf{r}_0$) along the great circle connecting $\mathbf{r}$ with the Sun position at time t. Actually it is $\mathbf{h}(t) = |\mathbf{r}_0 \times \mathbf{r}_s(t)|$, where $\mathbf{r}_s(t)$ is the Sun position vector.

   Let us consider two stars E_i and E_j separated by a measurable angle s_{ij} (Figure 3). The system of condition equations

$$- \sin \tilde{s}_{ij}\, \delta s_{ij} = \tilde{\mathbf{r}}_i(t) \cdot \delta \mathbf{r}_j(t) + \tilde{\mathbf{r}}_j(t) \cdot \delta \mathbf{r}_j(t), \qquad i, j = 1, \ldots, N;\tag{2}$$

where linearization is obtained by using approximate catalogue values ($\sim$), can be solved in principle by least square techniques to obtain positions, parallaxes and proper motions. The size of the normal system is $5N$ and the normal matrix is rank deficient due to the fact that only relative angular measurements are considered. Physically this results in 6 degrees of freedom of the reference system corresponding to an unknown rotation and its variation in time.

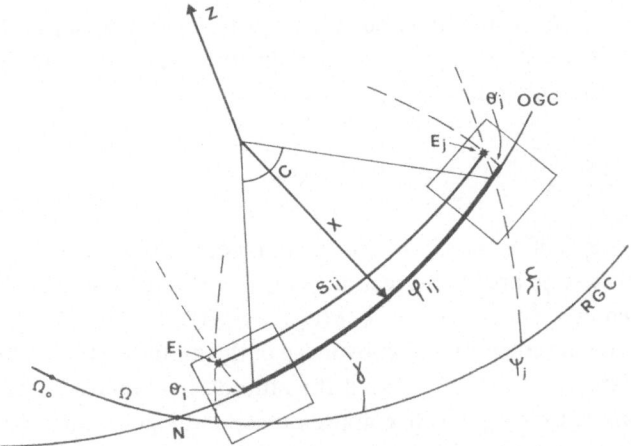

Fig. 3.   Geometry of angular distances measurement s_{ij} between stars E_i and E_j separated by a large angle.

The angular distances s_{ij} are measured by comparing in the focal plane of a single telescope the positions of the images of stars E_i and E_j belonging to two fields of view (FOV) separated by a large angle C (Figure 3), fields superposition beeing obtained by a complex mirror having two reflecting surfaces (Figure 4). A large value of C (between about 1 and $\pi - 1$ rad) is necessary to improve the accuracy of the arc measurements, to increase the rigidity of the reference system and to derive parallaxes by comparison of stars with different parallax factors. Let us consider a great circle OGC (Figure 3)

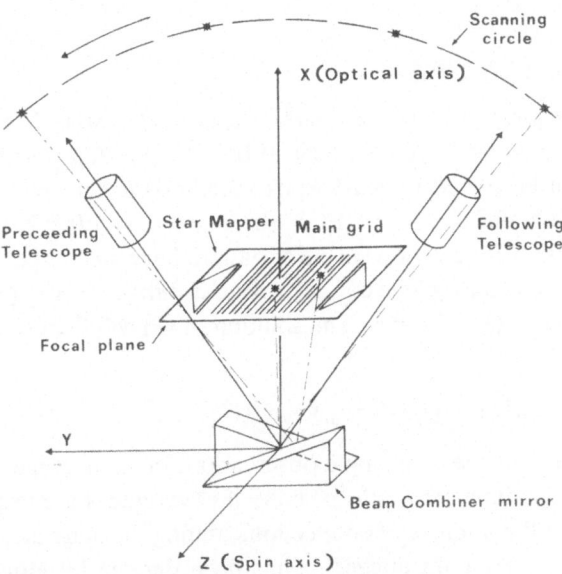

Fig. 4.   Concept of the HIPPARCOS mission.

close to stars E_i and E_j, which may be defined with respect to some Reference Great Circle (RGC) by its position angle γ and the longitude Ω of the node N, referred to as attitude parameters. The following variance estimate

$$\sigma^2(s_{ij}) = \frac{W^2}{6} \tan^2 \frac{C}{2} \ \sigma^2(\theta) + \sigma^2(\varphi_{ij}) , \tag{3}$$

holds if $2\theta = \theta_i + \theta_j$ is of order $1°$, $W/2$ is the maximum distance of $E_i(\theta_i)$ and of $E_j(\theta_j)$ from OGC and where φ_{ij} are the projections of s_{ij} onto OGC. It is seen that for C values in the range given above it is $\sigma(s_{ij}) \simeq 1.2\sigma(\varphi_{ij}) \simeq 0''001$ if $\sigma(\theta) \simeq 0''15$. This condition sets an accuracy requirement on the knowledge of the attitude θ (a combination of γ and Ω) of the OGC and it also shows that, if the attitude requirement is satisfied, we may measure only projections φ_{ij} and use approximate catalogue values for θ_i and θ_j. The angular distances φ_{ij} are obtained by rotating the telescope around an axis perpendicular to the OGC plane, defined in practice by the optical axes of the two viewing directions, at constant speed R (Figure 4). A grid of parallel opaque and transparent slits, placed on the focal plane of the telescope (not shown on the figure), modulates the light intensity from which the grid coordinate G of a star E at time t can be derived, reckoned, e.g., from the field center. The corresponding field coordinate $\eta = \eta(G)$ on the sky will depend on telescope distortions and focal plane imperfections. Assuming that the transformation $\eta(G)$ is known, it is represented by

$$\varphi_{ij} = \eta_j(t_j) - \eta_i(t_i) + R(t_j - t_i) + C . \tag{4}$$

Several connections φ_{ij} are possible among the $n = NW/2$ stars belonging to a given OGC, where W is now the FOV size (in radians), and we may build a system of condition equations

$$\varphi_{ij} - \tilde{\phi}_{ij} = d\varphi_i - d\varphi_j , \tag{5}$$

where φ_i and φ_j are longitudinal coordinates (abscissae) of stars E_i and E_j with respect to some arbitrary origin on the OGC, defined by, e.g., one star position or by $\Sigma\varphi_k = 0$. The system (5) can be solved by least square methods to improve $\tilde{\phi}_{ij}$.

At any instant on the combined FOV there are $m = NW^2/2\pi$ stars that can be observed. If successive observations of a star pattern have an overlap factor Q_f, the total number of observations during one complete scan is $2\pi mW^{-1}(1 - Q_f)^{-1}$ and each star will be observed $2(1 - Q_f)^{-1}$ times. The solution of (5) will then provide the following average variance

$$\sigma^2(\varphi_i) = 0.5(1 - Q_f) V\sigma^2(\eta_i) + \sigma^2(C)/2 , \tag{6}$$

where $\sigma^2(\eta_i)$ is the variance of a single observation for an average star. The factor V (V^{-1} is the rigidity parameter) would be unity if covariances are neglected. Actually it is a function of n, of the number of connections among the stars and of the basic angle $C \simeq 2\pi (g - 0.5)/n$, g being an integer. It has been derived by Hoyer et al. (1981) by solving an idealized system (5), assuming regular distribution of the n stars on the OGC scan and equal-observational weight to each star. For $20 < n < 2000$ and $C \simeq 1$ rad, it is

$$V \simeq 1 + 0.5n^{1/3}/(m - 1).\tag{7}$$

We see that rigidity will be improved by employing a large number of stars and a large field size W, which in turn would impose a larger accuracy on attitude determination. The attitude of the OGC scan will be derived by using a system of vertical and inclined slits (star mapper) placed on the focal plane of the telescope in front of the main grid (Figure 4).

Finally let N_{si} the number of OGC scans where star E_i has been observed for a mission duration T. By considering that each abscissa gives one coordinate out of two and that six observations per year are necessary to separate positions and proper motions, we have the following expected variances in units of $\sigma^2(\varphi_i)$

$$\sigma^2(\lambda_i) \simeq \sigma^2(\beta_i) = 2/N_{si},\tag{8}$$

$$\sigma^2(\mu_{\lambda_i}\cos\beta_i) \simeq \sigma^2(\mu_{\beta_i}) = 24/N_{si}T^2,\tag{9}$$

for positions λ_i and β_i and for proper motion components $\mu_{\lambda_i}\cos\beta_i$ and μ_{β_i}. Isotropic precision in each coordinate will be reached by random scanning. If the pole of OGC makes an angle ξ with the direction of the Sun and if u is the position angle of a star with respect to the Sun, as seen from the OGC pole, we have, in units of $\sigma^2(\varphi_i)$

$$\sigma^2(\omega) = 2N_{si}^{-1}\langle\sin^2\xi\sin^2 u\rangle^{-1},\tag{10}$$

where $\langle.\rangle$ indicates the average.

The final accuracy of astrometric parameters will then depend, beside a good performance of the telescope and the spatial modulator, on sky observing strategy, on attitude determination, on FOV size and on focal plane observing strategy. We also see that the angular standard of the mission, C, should be very stable over one complete scan, possibly to better than $0''.001$ (3σ).

3.2. THE PAYLOAD

The payload configuration is shown on Figure 5 and main parameters are given in Table II. The beam combiner, providing a fields splitting of $C = 58°$, is a flat aspheric mirror made of zerodur. It constitutes the entrance pupil (0.029 m^2) of an All Reflecting Schmidt-Telescope. The focal length is 1400 mm and the FOV size is $0.9 \times 0°.9$. The Schmidt corrector is shaped on the complex mirror itself and will have an elliptical profile because it works at high incidence ($14°.5$). Tolerances of optical manufacturing are set at $\lambda/60$ r.m.s. on the average. Detailed parameters of the several telescope elements are given, in the open literature, by Le Gall et al. (1983). The true value of the basic angle, to be accurately calibrated during the mission, should be stable to better than $0''.001$ (3σ) over very short time intervals (few hours) throughout the operational life of the satellite. This condition imposes severe constraints on thermal control performances in order to maintain temperature variations of the telescope mirrors below

P. L. BERNACCA

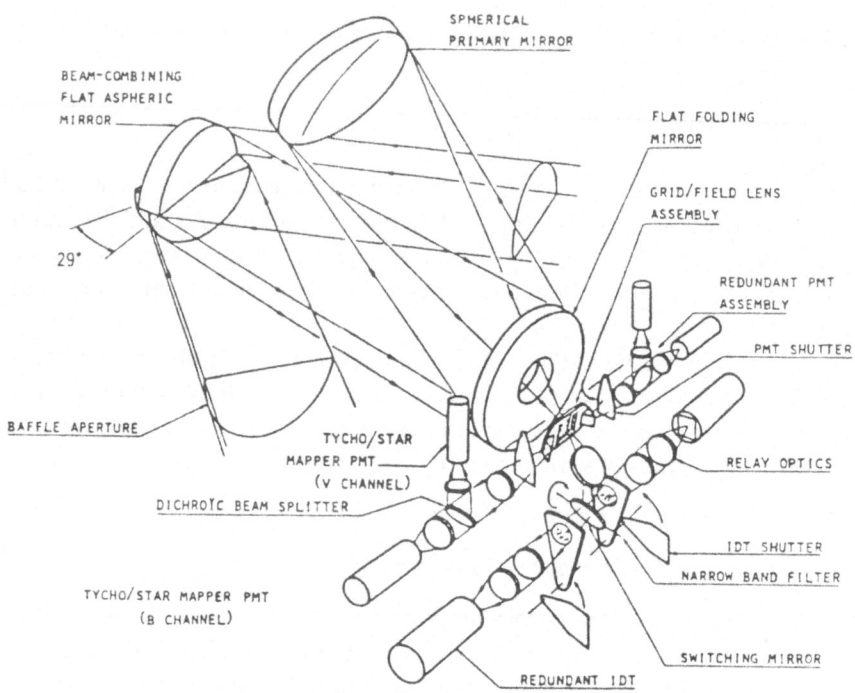

Fig. 5. Exposed view of HIPPARCOS payload.

0.2 K around 20 °C. The most severe constraint is control of the parabolic gradient variation of the complex mirror to better than 0.04 K.

Continous optical studies were performed since phase-A by systematically searching potentially attractive optical concepts. Specific points of concern were: (i) how to achieve the best FOVs splitting, i.e., the choice of the entrance pupil among double or triple split or semitransparent assemblies, (ii) wether dioptric front and field correctors could be allowed, (iii) elimination of chromatic aberration all over an unvignetted FOV and, (iv) assembly tolerances and manufacturing. A trace of these early studies is reported by Iorio-Fili and Scandone (1979) in the open literature. The following configurations were traded off: Ritchey–Chretien, Baker–Schmidt, Three Mirrors Systems, and All Reflecting Schmidt. The last one has proven to be the best. Nevertheless, because of the requirement of very high accuracy in image location, diffraction chromaticity is not negligible and can only be accounted for by calibration or by data reduction algorithms. This effect, which depends on the different repartition of energy in the diffraction pattern as a function of the star spectrum, produces a phase difference in the harmonics which describe the images of two stars located at the same position on the sky but having different colors. For the case of HIPPARCOS, diffraction chromaticity is a function only of the transversal coordinate on the focal plane and it may yield a positional error as large as 0″.0015 at the FOV edges for the nominal instrument (Le Gall *et al.*, 1983).

TABLE II

System parameters: HIPPARCOS

System parameter	Nominal value
– Revolving Scanning Angle	43°
– Number of revolutions of the spin axis	
about Sun direction per year	6.4
– Number of scans per day	11.25
– Transverse field of view	0.9°
– Longitudinal field of view	0.9°
– Basic angle	58°
– Grid period	1.208″
– Grid ratio	0.39
– Oversizing factor	1
– Magnification factor (IDT relay optics)	0.442
– Sampling period	$\frac{1}{1200}$ s
– IDT repositioning period	$\frac{1}{150}$ s
– Interlacing period	$\frac{2}{15}$ s
– Frame period	$\frac{32}{15}$ s (2.13 s)
– Entrance pupil diameter	290 mm
– Central obscuration ratio	0.38 (oblong)
– Focal length	1400 mm
– Field curvature radius	1400 mm (convex)
– Unprotected angles for straylight	
• lingitudinal	24°
• transverse	13.5°
– Maximum count rate	10 MHz
– IDT aperture diameter	90 μm

The two spatial modulators (main grid and star mapper) on the focal plane are implemented by electron beam techniques on a same curved substrate. The main grid consists of about 2700 parallel slits with period $s = 1″.2$. The transparent to opaque band width ratio is $d = 0.39$. The modulating grid is re-imaged onto the photocathode of an Image Dissector Tube (IDT) by means of a set of folding mirrors and relay lenses (Figure 5). Two relay lenses and two IDT are used in cold redundancy, the selection being obtained by a switching mirror.

The instananeous field of view (IFOV) of each IDT is nominally circular with a diameter of about $37″$, the sensitivity profile falling off rapidly outside, in order to isolate most of the star images from those of neighbour field stars or of angularly far physical companions. The IFOV position can be directed at any point of the FOV to within $5′$ by varying the currents applied to the deflection coils. The IDT converts the modulated light intensity into a sequence of photons counts which are sampled at a frequency of 1200 Hz and transmitted to the ground.

The grid period s has been chosen to maximize the ability of the instrument to detect double stars. Double star signals can be recognized only if the convolution of the diffraction pattern of a stellar source with the periodic grid transmittance can be described by at least two Fourier components, for, indeed, there is no way of decomposing back a purely sinusoidal signal into the signal of the two stars. Major contribution to double stars aspects at signal level are due to Lindegren (e.g., 1979). The oversizing factor (i.e., the ratio between the relay aperture and that of the entrance pupil) is set equal to unity, although the diffraction produced by the grid would require a large aperture to collect most of the photons. The adopted value, combined with the given grid ratio d gives a minimum in the photon statistical error of the positional information. This error depends, in fact, not only the steepness of the modulation curve (more steep for small d's) but also on the number of photons collected (more photons for large d's and oversizing factor).

The second modulator consists of two star mappers which are used in cold redundancy (Figure 4). Each star mapper emploies a star mapper grid located on the side of the main modulator and two photomultipliers (S20) which detect the light transmitted in two different spectral bands B_T and V_T, the spectral separation beeing done by means of a dichroic beam splitter (Figure 5). Each star mapper grid consists in four vertical and four inclined slits whose geometry is organised in such a way to allow the reconstitution of the satellite three-axes attitude from the analysis of the signal produced by the stars moving across. The modulated light intensity converted in photon counts by the PMT is sampled at a frequency of 600 Hz and transmitted to the ground. A study of high-accuracy star mappers and their use for spinning satellites is available from Carlucci and Donati (1981).

The payload includes also calibration and focusing devices, light baffles and shutters to prevent Earth-, Moon-, and Sunlight to enter the telescope. Shutters will be operated anytime the level of stray light exceeds a prefixed value. The unprotected FOV for straylight is defined in Table II.

The transmission curve (H-filter) of the payload for the case of observations through the main grid is compared with that of the *UBV* system on Figure 6 (Granès and Mignard, 1983). Absolute peak efficiency is 0.8% in the range 450 to 500 nm (MATRA,

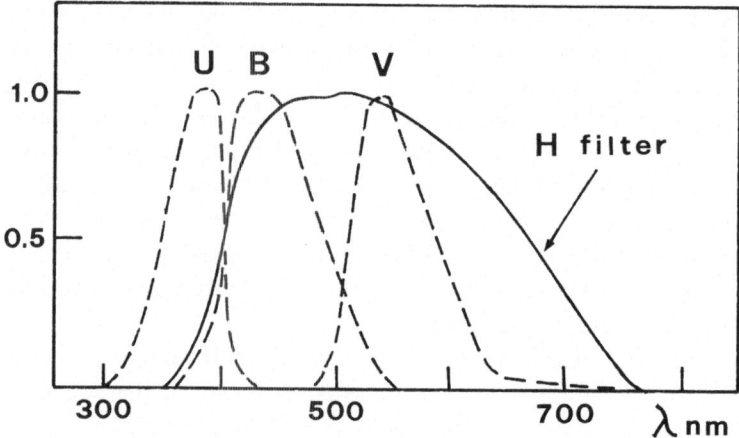

Fig. 6.   Comparison of the HIPPARCOS filter (*H*) with *UBV* system.

1984). The cut-off at short wavelengths minimizes possible effects induced by the steep gradient of the stellar flux around the Balmer discontinuity. It will be possible to convert *UBV* magnitudes to HIPPARCOS magnitudes H with the relation $H - V = 0.485(B - V) - 0.219(B - V)^2 + 0.022(B - V)^3$. The count rate (Hz) from a star of magnitude $B = 9$ is given in Table III (Le Gall and Gonin, 1984). The B_T channel of the star mapper detection chain has efficiencies of 2.5, 5.9, 6.3, 5.3, 2.9, and 1.2% at 375, 400, 420, 450, 475, and 500 nm, respectively. The V_T channel has efficiencies of 1.5, 3.9, 3.7, 2.7, 1.7, and 0.7% at 475, 500, 525, 550, 575, and 600 nm, respectively (MATRA, 1984). Outside of these spectral ranges, efficiencies decrease rapidly. Additional information on the star mapper is given in Section 3.

TABLE III

The intensity (ph/s) and the modulation factors for a star of magnitude
$B = 9$

$B - V$	I (Hz)	M1	M2	M3
− 0.25	1555	0.864	0.347	0.020
+ 0.50	1728	0.816	0.331	0.014
+ 1.25	3645	0.771	0.309	0.009

3.3. Observations and accuracy prediction

Homogeneous sky coverage is obtained by the so called revolving scanning, which is a combination of the following two motions (Lindegren, 1982).

– the satellite spins around the z-axis with constant angular velocity $R = 11.25$ revolutions per day.

– The z-axis moves around the Sun according to the law

$$v = \bar{v} - \cos \bar{v}/(K \tan \xi),$$

$$\bar{v} = \bar{v}_0 + K(\lambda_s - \lambda_{s0});$$

(11)

where v is the phase angle, λ_s is the Sun longitude, $K = 6.4$ is the number of revolutions per year, and $\xi = 43°$ is the angle between the direction of the z-axis and that of the Sun. Subscript '0' indicates initial conditions. The scanning law angles are illustrated on Figure 7.

The choice of the scanning law results from a trade-off between technical constraints and the implementation of the scientific requirements introduced in Section 3.1. Major driving consideration have included:

– Isotropic precision in positions requires that at every point in the sky there exist scanning great circles intersecting at right angles. This is possible with ξ close to 45°.

– For a certain mission length T there are certain discrete K-values which minimize the variation of the scan pattern with longitude.

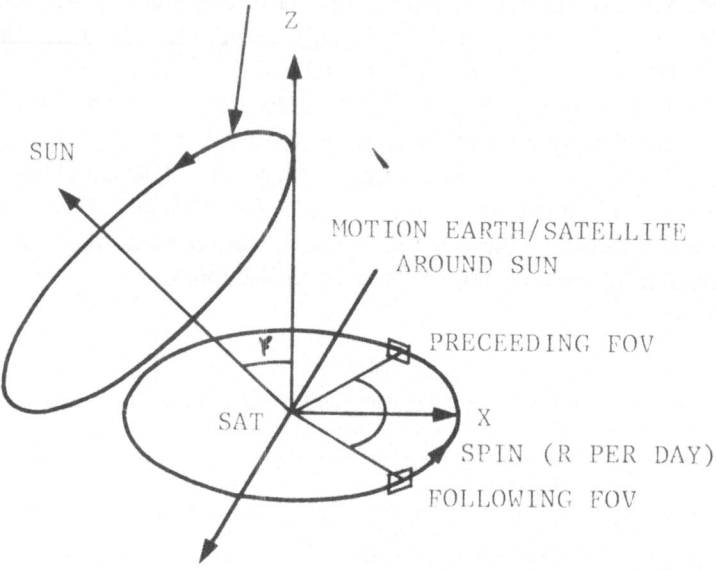

Fig. 7.   The revolving scanning law foreseen for **HIPPARCOS** mission.

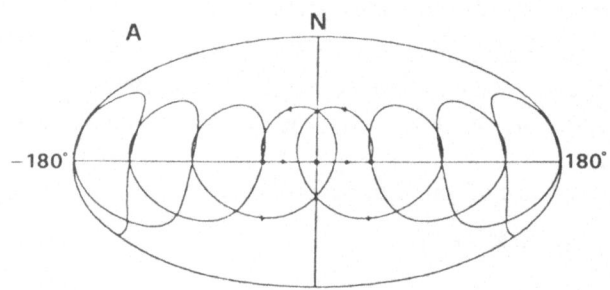

Fig. 8.   Loop of **HIPPARCOS** spin axis on the sky in ecliptic coordinates.

– For a given K there is a minimum revolving scanning angle $\xi = 270°/K$ which makes the loop of the z-axis on the sky to overlap with some margin, thus guaranteeing at least six observations per star every year. This is necessary for the separation of positions and proper motions. The loop of the z-axis is shown on Figure 8.

– The accuracy of parallax determination is improved if the projection of the parallactic displacement on a scanning circle is maximum, that occurs when $\xi = 90°$, the displacement occurring in the direction of the Sun.

– The fraction of interrupted scans due to occultations of the FOV by the Earth and

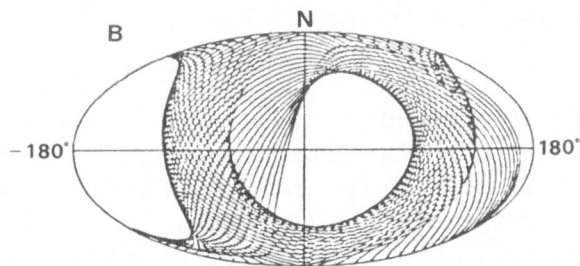

Fig. 9.   Scanning path of HIPPARCOS telescope during one loop of the spin axis.

Moon and, consequently, the dead time f of the mission because of intolerable stray light, decreases with ξ. With the present system parameters it is $f \simeq 0.08$.

 – Solar panels, thermal control and baffles design is easier with small ξ's.

The actual scanning path of the telescope optical axis during one loop of the spin axis is shown on Figure 9. The pattern can be translated along the ecliptic with period $2\pi/K$ in longitude to cover the sky in T years. Once a couple of values K and ξ are selected, the spin rate R can be obtained with the relation

$$V_z = RW(1 - Q_s),\qquad(12)$$

where W is the field size in degrees, Q_s is the overlap factor for two consecutive scans at antinodes and $V_z = \lambda_s(K^2 \sin^2 \xi + \cos^2 v)$ is the velocity of the spin axis on the sky. On the average it is $V_z = (1 + (K^2 - \frac{1}{2}) \sin^2 \xi)^{1/2}$, so that we may select R by assuming an average value of the overlap factor $Q_s = 0.56$. The chosen scanning law is not too far from providing a uniform (random) sky coverage and we may use the coefficients of improvements (8), (9), and (10) to roughly estimate the final accuracy on astrometric parameters, where $\langle N_{sl} \rangle$ is

$$N_s = 3.1874 T V_z (1 - f)(1 - Q_s)^{-1}.\qquad(13)$$

The total observing time planned for $N = 10^5$ stars is at least $T(1 - \alpha)(1 - f) = 6.5 \times 10^7$ s over $T = 2.5$ yr, where α stands for calibration time, commissioning and other time losses, so that the accuracy prediction in units of $\sigma(\varphi_i)$ is 0.17, 0.24, and 0.36 for positions, proper motions and parallax respectively. It is seen that $\sigma(\varphi_i) \simeq 0\rlap{.}''005$ is needed for final accuracies close to $0\rlap{.}''001$ r.m.s. Actual coefficients of improvements depend on the ecliptic latitude due to the variation of the scan pattern as visible on Figure 9 and, in addition, should be computed by examination of the scan pattern in detail (cf. Larcher, 1978; and Høyer *et al.*, 1981).

In order to see how such an abscissa accuracy is possible, let us consider the photoelectron counts recorded at time t_k and during the sampling interval $T_1 = \frac{1}{1200}$ s for the case of a single star image in the FOV. They can be modeled as a realization of a Poisson process with expected value

$$S_k = BT_1 + IT_1(1 + M_1 \cos(2\pi n_k + \phi_k) + M_2(4\pi n_k + 2\phi_k + v_c)), \qquad (14)$$

where n_k is the number of grid periods, M_1 and M_2, given in Table III, are the first two modulation coefficients, ϕ_k is the phase of the image photocenter, I is the stellar count rate (Hz) and B ($\simeq 50$ Hz) is the background intensity. The term v_c accounts for the phase difference due to diffraction chromaticity. If the attitude evolution is known with accuracy better than $0''.1$ r.m.s., then n_k is known without ambiguity and the estimation of $\phi(t)$ from the time series (14) will give the grid coordinate $G(t) = sn(t) + s\phi(t)/2\pi$ at some average instant t of the observations made at $t_k, t_{k+1}, \ldots$. Incidentally we may note that from (14) the parameters BT_1 and IT_1 can also be estimated.

The photon statistical error on the phase can be estimated by the Cramer–Rao bound

$$\sigma_p(\phi)^{-2} = \langle (\partial S_k / \partial \phi_k)^2 / S_k \rangle, \qquad (15)$$

where the average is computed over the observing interval τ. For $\tau = 1s$ and for the given M_1 and M_2 an approximation is $\sigma_p(\phi) \simeq 0.3 I^{-1/2}$ arcseconds. Other sources of error per second of observing time include $\sim 0''.003$ r.m.s. due to IDT photocathode inhomogeneities and grid irregularities and $\sim 0''.002$ r.m.s. due to vibrations (jitter) induced by solar panels oscillation after the actuation of gas jets, which are used to control the satellite attitude. In addition, some 25% of $\sigma_p(\phi)$ should be attributed to IFOV depointing and other minor effects. As far as chromaticity is concerned, the residual error on $G(\eta)$ for stars of color index $B - V = 1.25$ and $B - V = -0.25$ with respect to the grid coordinate of a $B - V = 0.5$ star having the same η, is estimated of about 2.5 milliarcsecond (mas) at field edges, with a view to expected instrumental deformations enhancing the nominal effect. If this constant chromaticity will be calibrated, a r.m.s. residual error of 1.5 mas will probably be at work in the data at all $B - V$. By combining the above error sources we have $\sigma(\phi) \simeq 10$ mas per second of observing time for a star of magnitude $B = 9$. If two stars of magnitudes m and m' are given observing times τ and τ', the variance of an angle measurement φ_{ij} is minimized by the partitioning $\tau/\tau' = \sigma_p(m, \tau = 1s)/\sigma_p(m', \tau' = 1s)$. The sharing of the planned observing time among all programme stars is called global observing strategy and it is given in Table IV.

Actual observations will be performed in the following way. The on board computer, using preliminary attitude knowledge from gyroscopes readings and star mapper observations, computes the times of transit of the programme stars into the main FOV and their successive positions on it. The IDT IFOV is then piloted to detect the selected stars in an observing sequence, called local observing strategy, which is described by four fixed periods as follows. Photoelectrons from a star will be counted every $\frac{1}{1200}$ s (sampling period T_1) for an uninterrupted duration of 1 grid period (IDT repositioning period $T_2 = 8T_1$). The IFOV will then be moved to the new position of the same star or to that of another star of the selected pattern. There will be 20 of these T_2 periods to be shared by all stars in the pattern according to a dwell time strategy to be implemented (cf. Kovalevsky and Dumoulin, 1983). Afterwards, the same star pattern is reobserved identically 16 times to obtain a frame of observations of $\frac{32}{15}$ s duration.

TABLE IV

Observing strategy (headings are explained at the bottom of the table)

B	N_B	n_B (OGC)	τ_B	τ_B (OGC)	$n(T_2)$
<6	3 000	24	270	4.0	2
6–7	5 400	42	310	4.6	3
7–8	14 800	116	390	5.9	4
8–9	40 800	320	530	8.0	5
9–10	16 000	126	730	11.0	7
10–11	12 000	94	990	14.9	9
11–12	6 000	47	1330	20.0	12
12–13	2 000	16	1780	26.7	16
Total	100 000	785	6510	7680	–

B: magnitude of stars; N_B: number of programme stars; n_B (OGC): number of stars on a great circle scan; τ_B: total observing time per star; τ_B (OGC): observing time per star per scan; $n(T_2)$: number of elementary periods T_2 of observations during each interlacing period $T_3 = 20T_2$.

Since a star will cross the FOV in 19.2 s, it results $1 - Q_f = \frac{1}{9}$. For our $B = 9$ star, observed about 0.6 s per frame it is $\sigma(\eta) \simeq 13$ mas if we consider a residual r.m.s. per frame of about $0''\!.003$ when applying the $\eta(G)$ transformation, in addition to the grid coordinate error. If a $B = 9$ star is an average star on the OGC scan, we finally have $\sigma(\varphi) \simeq 4.8$ mas. Actual accuracy assessment should take into account the relative weight of all measurements of a certain star with respect to the average observational weight defined by all stars belonging to the great circle scan.

Table V shows the expected accuracies for all astrometric parameters derived during the detailed study phase (Bouffard and Zeis, 1983) at $B = 9$ and $B = 12$, averaged all over the sky.

It should be noted that the scanning path of the optical axis deviates continuously from a great circle due to the motion of the spin axis. At the central instant of each frame an osculating great circle (OGC) can be defined by attitude parameters and the angles φ_{ij} can be projected onto a fixed RGC (Figure 3). If $\gamma \simeq 1°$, the attitude error $\sigma(\theta) \simeq 0''\!.1$ will induce a r.m.s. error $\langle (\zeta_i - \zeta_j)^2 \rangle^{1/2} \sigma(\theta) \simeq 0.6$ mas on the projections ψ_{ij}, ζ_i and ζ_j beeing the distances between stars E_i and E_j and the RGC. It is actually on the RGC that abscissae ψ_i are defined in the data reduction procedure.

3.4. THE TYCHO PROGRAMME

The star mapper (SM), called TYCHO, provides calibration data for the real time three-axes attitude determination needed for proper IDT piloting. The requirements for this purpose on accuracy and frequency of the observations are not much demanding: about $1''$ r.m.s. for a few stars distributed along the great circle scan. By interpolating gyroscopes readings and transit times of these stars through the SM central line, the spin

TABLE V

Predicted accuracies for the HIPPARCOS mission

Magnitude Color		$B = 9$			$B = 12$		
		$B - V = -0.25$	$B - V = 0.5$	$B - V = 1.25$	$B - V = -0.25$	$B - V = 0.5$	$B - V = 1.25$
Frame r.m.s. error		13.3 mas	11.6 mas	9.43 mas	39.3 mas	32.4 mas	24.4 mas
Star abscissa r.m.s. error		6.27 mas	5.97 mas	5.81 mas	11.5 mas	10.0 mas	8.46 mas
Astrometric parameter r.m.s. error	$\sigma a1$	1.51 mas	1.27 mas	1.42 mas	2.58 mas	2.16 mas	1.95 mas
	$\sigma a2$	1.98 mas yr^{-1}	1.79 mas yr^{-1}	1.85 mas yr^{-1}	3.53 mas yr^{-1}	3.02 mas yr^{-1}	2.62 mas yr^{-1}
	$\sigma a3$	1.20 mas	1.04 mas yr^{-1}	1.13 mas	2.08 mas	1.75 mas	1.56 mas
	$\sigma a4$	1.69 mas yr^{-1}	1.48 mas yr^{-1}	1.59 mas yr^{-1}	2.95 mas yr^{-1}	2.50 mas yr^{-1}	2.21 mas yr^{-1}
	$\sigma a5$	1.75 mas yr^{-1}	1.61 mas yr^{-1}	1.63 mas yr^{-1}	3.16 mas	2.71 mas	2.34 mas
Specification σai $i = 1, \ldots, 5$		2.60 mas yr^{-1}	2.00 mas yr^{-1}	2.60 mas yr^{-1}	5.20 mas yr^{-1}	4.00 mas yr^{-1}	5.90 mas yr^{-1}

$\sigma a1$: longitude; $\sigma a2$: proper motion in longitude; $\sigma a3$: latitude; $\sigma a4$: proper motion in latitude; $\sigma a5$: parallax.

velocity R, or $v = 168''75\ \text{s}^{-1}$ will be determined by the on board computer with an uncertainty of about 2%. The TYCHO–SM has however the intrinsic capability of providing positional information to better than $0''1$ r.m.s. which is needed for on-ground attitude reconstitution. The SM geometry is shown in Figure 10, where s^* is the period.

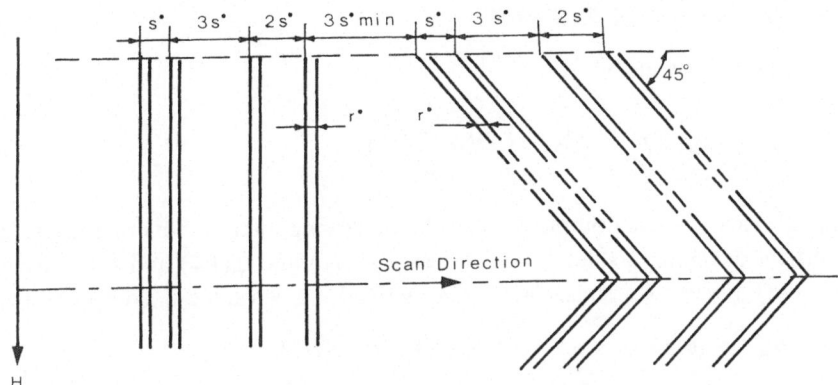

Fig. 10.   Geometry of TYCHO star mapper on board of HIPPARCOS.

SM parameters are given in Table VI (MATRA, 1984). The SM counts S_k recorded at time t_k in each B_T or V_T channel can be modelized by the Poisson process, of expected value $S_k - \varepsilon_k$, ε_k beeing the stochastic deviation, given by

$$S_k = B\tau + I\tau \sum_1^m{}_j f(G_k - G_j) + \varepsilon_k\,,\tag{16}$$

where G_j is the position of the median of the jth slit with respect to the SM center line, τ is the sampling interval, G_k is the SM grid coordinate of the image photocenter at time t_k, $f(.)$ is the slit spread function which describes the optical plus detector response.

TABLE VI

System parameters: TYCHO

System parameter	Nominal value
Grid period	$5.625''$
Vertical slit width	$0.90''$
Inclined slit width	$0.90''$
Slit height	$40'$
Number slits	4
Inclined slit inclination	$45°$
Grid pattern signature	83
SM sampling period	$(\frac{1}{600})$ s
Maximum count rate	4.9 MHz

The functions $f(.)$ are hump-type response and are available from Le Gall and Saisse (1983) for each channel, each type of slit and each color. For uniform scanning speed v we can write $f(G_k - G_j) = f(kv\tau - \varphi - G_j)$, where $\varphi = v(t_c - t_0)$, t_c being the transit time through the SM center line and t_0 a fixed initial time. An estimation $\hat{\varphi}$ of the SM phase φ follows from

$$\hat{\varphi} = \arg \max_\varphi \Sigma_k \Sigma_j S_k f(kv\tau - \varphi - G_j) ; \tag{17}$$

and the variance is given by

$$\sigma_\varphi^{-2} = \Sigma_k \langle S_k - \varepsilon_k \rangle^{-1} \left(\frac{\partial (S_k - \varepsilon_k)}{\partial \varphi} \right)^2 , \tag{18}$$

where the average $\langle . \rangle$ is computed over the observing interval covering the group of $m = 4$ vertical or inclined slits. By considering $\varphi(B_T)$ and $\varphi(V_T)$ and their respective variances (18) a weighted phase φ_0 can be derived. Its variance is computed from

$$\sigma_{\varphi_0}^{-2} = (\sigma_\varphi^2(V_T) + \sigma_\varphi^2(B_T))(\sigma_\varphi^{-2}(V_T)\sigma_\varphi^{-2}(B_T)) \tag{19}$$

and the positional error perpendicular to the scan is

$$\sigma_H^2 = \sigma_{\varphi_0}^2 (0°) + \sigma_{\varphi_0}^2 (45°) . \tag{20}$$

Preliminary values of $\sigma_{\varphi_0}^2$ as derived by simultation (Civalleri and Canuto, 1984) are given in Table VII for a single crossing of all vertical and (in brackets) inclined slits. Values of B and I are also given for different color indices. SM grid coordinates should be transformed in field coordinates η and ζ through the $\eta(G)$ transformation extrapolated from the main FOV and corrected for chromaticity when $(B - V) \neq 0.5$. For an expected $0{.}''025$ r.m.s. error on the transformation and few mas r.m.s. error due to chromaticity, we see that the positional error on η and ζ is less than $0{.}''1$ r.m.s. as requested. Of course these represent final accuracy to be fully exploited by the off-line attitude reconstitution.

It has been repeatedly pointed out by Høg in a series of notes (e.g., Høg et al., 1982) that this positional information should be exploited by a TYCHO Programme aimed

TABLE VII

Star mapper phase error for single crossing of four vertical slits (or four inclined slits)

Magnitude B	$I(B - V = 0.0)$ (Hz)		σ_{φ_0} (mas)		
	V_T	B_T	$B - V = 0$	$B - V = 0.5$	$B - V = 1.5$
7.0	21000	48000	6(8)	7(9)	7(10)
8.0	8300	19000	10(14)	12(16)	12(17)
9.0	3300	7600	17(26)	21(31)	22(33)
9.5	2100	4800	24(36)	28(44)	31(47)
10.0	1300	3000	32(52)	40(65)	43(69)
10.5	830	1900	46(76)	57(97)	62(104)

Background: $V_T = 1820$ Hz; $B_T = 2310$ Hz.

at obtaining a catalogue of many more stars than foreseen by the HIPPARCOS mission. During the mission a star will cross the SM about 50 times. However, the number of useful observations, in addition to those planned for the main mission, will be limited by the telemetry rate. It has been estimated that 16 to 32 observations could become available for 400 000 stars brighter than $B = 11$. The goal is to reach an imprecision of the order of about 30 mas r.m.s. in each coordinate. The transmission of all SM data to the ground and off-line detection of stars in the continuous photon counts record would provide positional information for additional 800 000 stars in the magnitude range $B = 11$ to $B = 12$ with an error of about $0''.15$ r.m.s.

The TYCHO programme includes also a magnitudes catalogues. A preliminary estimate $\hat{a}$ and $\hat{b}$ of $a = I\tau$ and $b = B\tau$ for each channel can be derived from

$$\text{Min}_{(a, b)} \Sigma_k (S_k - b - a \Sigma_j (kv\tau - \hat{\varphi} - G_j))^2 \, , \tag{21}$$

so that the Poissonian variance $\sigma^2(S_k) = \hat{b} + \hat{a}f(.)$ provides the final estimation $\hat{a}$ and $\hat{b}$ by means of the Gauss–Markov procedure

$$\text{Min}_{(a, b)} \Sigma_k \sigma(S_k)^{-2} (S_k - b - a \Sigma_j f(.))^2 \, . \tag{22}$$

The error on $\hat{a}$ and $\hat{b}$ follows from the covariance matrix of the process. A simple estimate of the error on the magnitude is possible if the background can be accurately

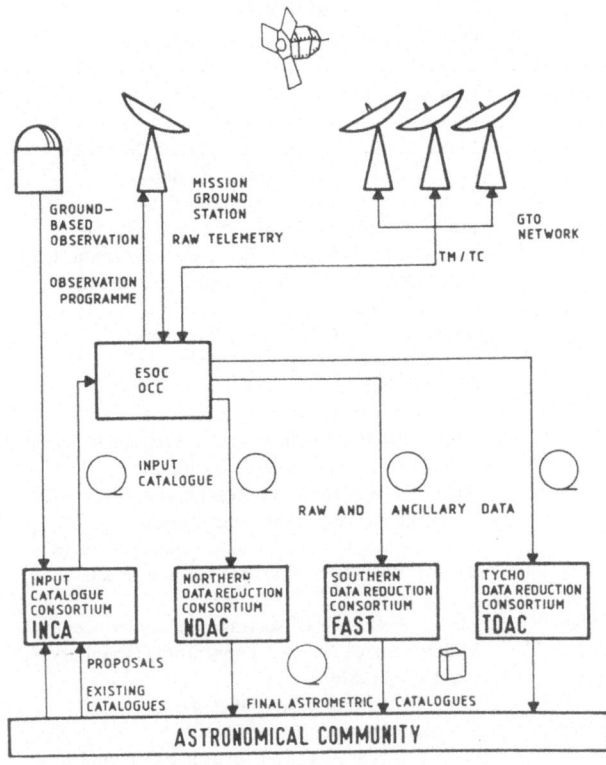

Fig. 11.   Organisation of the HIPPARCOS mission.

measured while no star is in the slits. For a single crossing of m slits we have

$$\sigma^2(S)S^{-2} = v(1 + I/B)(mIr)^{-1} \,, \tag{23}$$

where r is the slit width. The goal is to reach 0.2 to 0.3 mag. imprecision for stars between about $B = 10$ and $B = 11$ by combining all the observations at the end of the mission.

4. Mission Organisation

Figure 11 illustrates the system employed to achieve the scientific objectives (ESA, 1983). The programme of observations will be based on an Input Catalogue (INCA) containing all the necessary information on the programme stars like available positions and proper motions, provisional parallaxes even for the faintest stars, radial velocities, magnitudes and colors when available. All the available information on multiplicity will also be included. INCA is beeing prepared by a Consortium of scientific Institutes on the basis of research proposals submitted to ESA in response to an Announcement of Opportunity. Some 600 000 stars have been proposed for HIPPARCOS by astronomers all over the world. The final list will probably contain about 115 000 stars. A number of observatories are collaborating to determine positions and magnitudes for those stars which do not have such an information. The network of the programme stars will be

TABLE VIII

Main mission data reduction consortia

Consortium and leader	Leading institutes
F.A.S.T.: Fundamental Astronomy by Space Techniques (Southern Data Reduction Consortium) Leader: Prof. J. Kovalevsky, CERGA (Grasse)	Laboratoire d'Astronomie Spatiale, Marseille Centro di Studi sui Sistemi, Torino Dipartimento di Astronomia di Padova Dept. of Geodesy, Delft Univ. of Technology Istituto Matematico S. Pincherle Università di Bologna Astronomisches Recheninstitut, Heidelberg CERGA, Grasse Jet Propulsion Laboratory, Pasadena Space Research Laboratory, Utrecht CNES, Toulouse Bureau des Longitudes, Paris Centro Studi e Applicazioni in Tecnologie Avanzate, Bari Istituto Nazionale di Ottica, Firenze Istituto di Topografia, Fotogrammatria e Geofisica, Milano Osservatorio Astronomico di Torino Sterrekundig Institut Rijksuniversiteit Utrecht Istituto Astrofisica Spaziale CNR-Frascati
N.D.A.C.: Northern Data Analysis Consortium Leader: Dr. E. Høg, Copenhagen University Observatory	Royal Greenwich Observatory, Herstmonceux Mullard Space Science Laboratory, Holmbury St. Mary, Surrey Lund Observatory Copenhagen University Observatory Geodetic Institute, Copenhagen Danish Space Research Institute, Copenhagen

sufficiently regular so that the frequency of instances with too few or too many stars inside the FOV is minimized.

The IDT control process at OCC (ESOC, Darmstadt) identifies those stars which will be in the telescope field of view during successive intervals (30 m) on the basis of the nominal scanning law and it updates the on board IDT piloting sequence using pre-selected local and dwell time strategies. When the attitude of the satellite will deviate more than 10′ from the nominal scan, because of perturbations on the satellite dynamics, it will be up to the on-board computer to actuate gas jets (it will occur every 400 s on the average) to keep the scan close to the selected path. The following major data streams will be generated by the satellite: photon counts, gyro data, star-mapper data, house-keeping data, and calibration information and real time attitude determination. These data will be transmitted to ESOC via the Odenwald Ground Station and, after some pre-processing, will be distributed to the Data Reduction Consortia. The scientific data processing for TYCHO will be performed by the TYCHO Consortium which will utilize the data generated continuously by the star mapper. The measurements in the star mapper data stream will be correlated with real stars using a catalogue which will be derived from the *Space Telescope Guide Star Catalogue*.

TABLE IX

Input Catalogue and TYCHO Consortia

Consortium and leader	Leading institutes
I.N.C.A.: Input Catalogue Consortium Leader: Dr. C. Turon-Lacarrieu Obs. de Paris, Meudon	Sterrenwacht, Leiden Observatoire de Paris-Meudon Observatoire Royal de Belgique, Bruxelles Astronomisches Recheninstitut, Heidelberg Royal Greenwich Observatory, Herstmonceux Universidad de Barcelona Observatoire de Genève Centre de Données Stellaires, Strasbourg Observatoire de Bordeaux Observatoire de Besançon
T.D.A.C.: Tycho Data Analysis Consortium Leader: Prof. M. Grewing, Univ. Tübingen	Department of Astronomy, Padova Centro Studi sui Sistemi, Torino Centre de Données Stellaires, Strasbourg Institut für Astronomie, Tübingen Copenhagen University Observatory Danish Space Research Institute, Copenhagen Space Research Laboratory, Utrecht CERGA, Grasse Laboratoire d'Astronomie Spatiale, Marseille Lund Observatory Royal Greenwich Observatory, Herstmonceux University of London Observatory Huygens Laboratory, Leiden Space Telescope Science Institute, Baltimore Laboratorium voor Ruimteonderzoek, Groningen

The attitude reconstitution from the star mapper transit times and the processing of science data for the HIPPARCOS mission will be performed, independently, by two Consortia named FAST and NDAC. Summary information on the four Consortia is given in Tables VIII and IX. The methods for the preparation of the Input Catalogue and the algorythms to be implemented for data reduction have been proposed to ESA in 1981 in response to Announcements of Opportunities by all Consortia. In the open literature they can be found in the Proceedings of the FAST Thinkshop held in Asiago in June 1983 (Bernacca, 1983). A summary description is available in a review paper on Space Astrometry by Kovalevsky (1984).

Acknowledgement

The Steering Committee of FAST Consortium has agreed upon utilization of unpublished notes and working documentation for review papers on HIPPARCOS by members of the Consortium. Italian participation to FAST Consortium is funded by Piano Spaziale Nazionale of C.N.R.

References

Bacchus, P. and Lacroute, P.: 1974, in W. Gliese, C. A. Murray, and R. H. Tucker (eds.), 'New Problems in Astrometry', *IAU Symp.* **61**, 27.

Barbieri, C. and Bernacca, P. L.: 1979, *Colloquium on European Satellite Astrometry*, University of Padova.

Bernacca, P. L. and Cornelio, F.: 1978, *Astrometry Satellite Phase-A Study*, Aeritalia Final Report to ESA, Vols. 1 and 2.

Bernacca, P. L.: 1983, *The FAST Thinkshop*, University of Padova.

Boissières, J. R. and Larcher, S. J.: 1978, *Astrometry Satellite*, Theoretical Study of the Accuracy, Vol. 2, Application Mathématiques et Logiciels, Paris.

Bouffard, M. and Zeis, E.: 1983, in P. L. Bernacca (ed.), *The Fast Thinkshop*, University of Padova, p. 31.

Carlucci, D. and Donati, F.: 1981, *Proceedings of the 8th IFAC World Congress* **16**, 64.

Civalleri, P. P. and Canuto, E.: 1984, *HIPPARCOS Working Report CSS-2210/2*, Centro di Studi sui Sistemi, Turin.

Dieckvoss, W., Kox, H., Gunter, A., and Brosterhus, E.: 1975, AGK3, Hamburg, Bergedorf.

Duncombe, R. L., Benedict, G. F., Hemenway, P. D., Jefferys, W. H., and Shelus, P. J.: 1982, *The Space Telescope Observatory*, NASA CP-2244, 114.

ESA: 1977, *Report on the Mission Definition Study*, ESA-DP/PS(76)11, Rev. 1.

ESA: 1979a, *HIPPARCOS Space Astrometry*, ESA-SCI(79)10.

ESA: 1979b, *HIPPARCOS Spacecraft Design and Development*, ESA-ESTEC, PF616.

ESA: 1983, *Mission Implementation Requirements*, ESA-HIP-1035.

Fanselow, J. L., Sovers, O. J., Thomas, J. B., Cohen, E. J., Purcell, G. H., Rogstad, D. H., Skierve, L. J., and Spitzmesser, D. J.: 1984, *Astron. J.* **89**, 987.

Fricke, W.: 1977, Veröff. Astron. Rechen Institut, Heidelberg, No. 28.

Fricke, W.: 1982, in M. A. C. Perryman and T. D. Guyenne (eds.), *Proceedings of an International Colloquium on the Scientific Aspects of the HIPPARCOS Mission*, ESA SP-177, 43.

Fricke, W. and Koppf, A.: 1963, *Fourth Fundamental Catalogue (FK4)*, Veröff. Astron. Rechen Institut Heidelberg, No. 10.

Granès, P., Mignard, F., 1983, in P. L. Bernacca (ed.), *The FAST Thinkshop*, University of Padova, p. 99.

Hoyer, P., Poder, K., Lindegren, L., and Høg, E.: 1981, *Astron. Astrophys.* **101**, 228.

Høg, E., Jaschek, C., and Lindegren, L.: 1982, in M. A. C. Perryman and T. D. Guyenne (eds.), *Proceedings of an International Colloquium on the Scientific Aspects of the HIPPARCOS Mission*, ESA SP-177, 21.

Hughes, D. W.: 1984, *Nature* **307**, 15.

Iorio-Fili, D. and Scandone, F.: 1979, in C. Barbieri and P. L. Bernacca (eds.), *Colloquium on European Satellite Astrometry*, University of Padova, p. 29.

Kovalevsky, J.: 1975, Space Astrometry, ESRO Symposium, Frascati, 22–23 October, 1974, 67.

Kovalevsky, J.: 1984, *Space Sci. Rev.* (in press).

Kovalevsky, J. and Dumoulin, M. T.: 1983, in P. L. Bernacca (ed.), *The FAST Thinkshop*, University of Padova, p. 271.

Lacroute, P.: 1968, in L. Perek (ed.), *Transactions of the IAU* **XIIIB**, 63.

Larcher, S. J.: 1978, *Astrometry Satellite, Theoretical Study of the Accuracy*, Vol. 1, Application Mathematiques et Logiciels, Paris.

Le Gall, J. Y. and Saisse, M.: 1983, LA-YY-02, issue 1, Laboratoire d'Astronomie Spatiale, Marseille.

Le Gall, J. Y. and Gonin, J. C.: 1984, LAS/FAST/26, issue 1, Laboratoire d'Astronomie Spatiale, Marseille.

Le Gall, J. Y., Saisse, M., Tayeb, C., Froeschle, M., and Mignard, F.: 1983, in P. L. Bernacca (ed.), *The FAST Thinkshop*, University of Padova, p. 51.

Lindegren, L.: 1979, in C. Barbieri and P. L. Bernacca (eds.), *Colloquium on European Satellite Astrometry*, University of Padova, p. 15.

Lindegren, L.: 1982, Unpublished Note, 19 May.

MATRA: 1984, *HIPPARCOS Payload System Specifications*, SY.4.00.0.

Perryman, M. A. C. and Guyenne, T. D.: 1982, *Proceedings of an International Colloquium on The Scientific Aspects of the HIPPARCOS Mission*, ESA-SP-177.

Preston, R. A., Lestrade, J.-F., and Mutel, R. L.: 1983, in P. L. Bernacca (ed.), *The FAST Thinkshop*, University of Padova, p. 395.

Roeser, S.: 1983, in P. L. Bernacca (ed.), *The FAST Thinkshop*, University of Padova, p. 391.

Soederhjelm, S. and Lindegren, L.: 1982, *Astron. Astrophys.* **110**, 156.

Westerhout, G. and Hughes, J. A.: 1982, in M. A. C. Perryman and T. D. Guyenne (eds.), *Proceedings of an International Colloquium on the Scientific Aspects of the HIPPARCOS Mission*, ESA-SP-177, 69.

HIPPARCOS ASTROMETRIC BINARIES*

J. DOMMANGET

Royal Observatory, Belgium

(Received 30 July, 1984)

Abstract. The Input Catalogue of some 100 000 stars that is presently prepared for observation by the astrometric satellite HIPPARCOS, will contain many double and multiple systems. Because of the Hipparcos observation technique, these systems have to be divided in a few particular categories that are described and discussed. Each of them leads to specific considerations concerning the contribution of the Hipparcos observations. The category of very close pairs to which Hipparcos will certainly add many systems newly discovered during the mission, is compared to that of the few astrometric pairs that have been discovered by groundbased techniques.

Hipparcos appears finally as a very important tool in double star astronomy research and especially in the field of very close systems.

1. The Reasons of an Invited Talk

The reasons of having some talks devoted to the astrometric satellite HIPPARCOS at the occasion of an International Conference commemorating Bessel's second centenary of birth appear evident. From a rapid examination of the amount of basic information that will be collected by HIPPARCOS, it turns out that this space astrometry mission lies in a perfect continuation of the researches that F. W. Bessel inaugurated or fundamentally developed. Establishment of stellar catalogues and determination of parallaxes are two of the main features that immediately should be recalled when one has in mind Bessel's work and the fundamental aims of the Hipparcos satellite.

But when the conference is limited to 'astrometric binaries' the link between Bessel's research on Sirius and Procyon on one side, and the Hipparcos mission on the other, appears less evident and it is understandable that one should wish some more details about how this satellite may be an original tool for some progress in that field.

It will be my duty to try to show here how far this may be the case.

2. Double Star Observation by Hipparcos

In his invited talk, Dr P. L. Bernacca has just been fully describing the Hipparcos project. I have thus not to come back to the Hipparcos technical observational principle. In a paper given at the *IAU Colloq.* No. 62, held at Flagstaff in May 1981 (Dommanget, 1983a), I have shown what double star astronomy may expect from the Hipparcos astrometry mission. A summary has been presented at the International Colloquium held at Strasbourg by ESA in February 1982 (Dommanget, 1982b). I will, therefore, only

---

* Communication presented at the International Conference on 'Astrometric Binaries', held on 13–15 June, 1984, at the Remeis-Sternwarte Bamberg, Germany, to commemorate the 200th anniversary of the birth of Friedrich Wilhelm Bessel (1784–1846).

Astrophysics and Space Science **110** (1985) 47–63. 0004–640X/85.15
© 1985 *by D. Reidel Publishing Company.*

recall here the basic problems emerging from the existence of binaries in the *Hipparcos Input Catalogue* and, consequently will try to show the place of astrometric binaries in the whole project.

We know that because of the principle of the Hipparcos scanning technique and observational strategy, the catalogue of all stars to observe (100 000) should be ready at launch*. And because of the technical features related to the use of an Image Dissector Tube (IDT) and the transmission characteristics inside its Instantaneous Field of View (IFOV), the position of each star should be given with an accuracy at least of the order of $\pm 1''$. All these considerations are based on the assumption that each star is a single one. As soon as the star is a double or a multiple system problems of a different nature appear. It is clear that when the apparent separation of their components is much greater than the size of the IFOV (very wide pairs), each component is separately observable as a single star. When at the contrary, they are as close as making the Hipparcos technique unable to discover any duplicity (very close spectroscopic pairs), the entire system will be observed as a single star. But for intermediate separations, different situations will appear and should be distinguished.

To better pose the problem of binaries in the Hipparcos mission, it appears suitable to consider first some particular values of the separation of their components from the point of view of double star astronomy as well as from that of the Hipparcos observational technique.

These values – which principally concern stars of intermediate values of Δm – are given on Figure 2 and are hereunder commented on:

(a) $\rho = 20''$ is, following Bacchus (1982, 1984), the minimum value that should exist between the two components of a double star, for making their observation individually possible. For somewhat shorter separation, it appears impossible to center the IFOV in any way on one of both stars without receiving on the photocathode of the IDT, some disturbing energy originating from the other; this would lead to no significant analysis of the photometric signal;

(b) $\rho = 5''$ and $2''$ are limits also established by Bacchus (1982, 1984). The first one is a maximum beyond which a sensible loss of energy on the photocathode from one or from both stars will occur, whatsoever the centering of the 'system' may be in the IFOV. The $2''$ limit is found by considering also the transmission profile of the IFOV as expected from recent ESA information and the 'veiling glare' effect (Figure 1), with the condition that the centering of the system is made possible with an accuracy of the order of $\pm 1''$;

(c) $\rho = 0\rlap{.}''6$ is the half-slit separation of the modulation grid as well as the resolving power of the Hipparcos complex mirror in the scanning direction (25 cm broadness);

(d) $\rho = 0\rlap{.}''2/0\rlap{.}''1$ is the limit of astrometric observability of a binary (resolved components) by the Hipparcos technique for each favourable scan, as Bacchus (1982) has

---

* A Catalogue of the Components of Double and Multiple Stars (C.C.D.M.) is presently prepared on the basis of the *Index Catalogue* and personal documentation, in order to furnish all needed information on systems appearing in the *Hipparcos Input Catalogue* (Dommanget, 1983b).

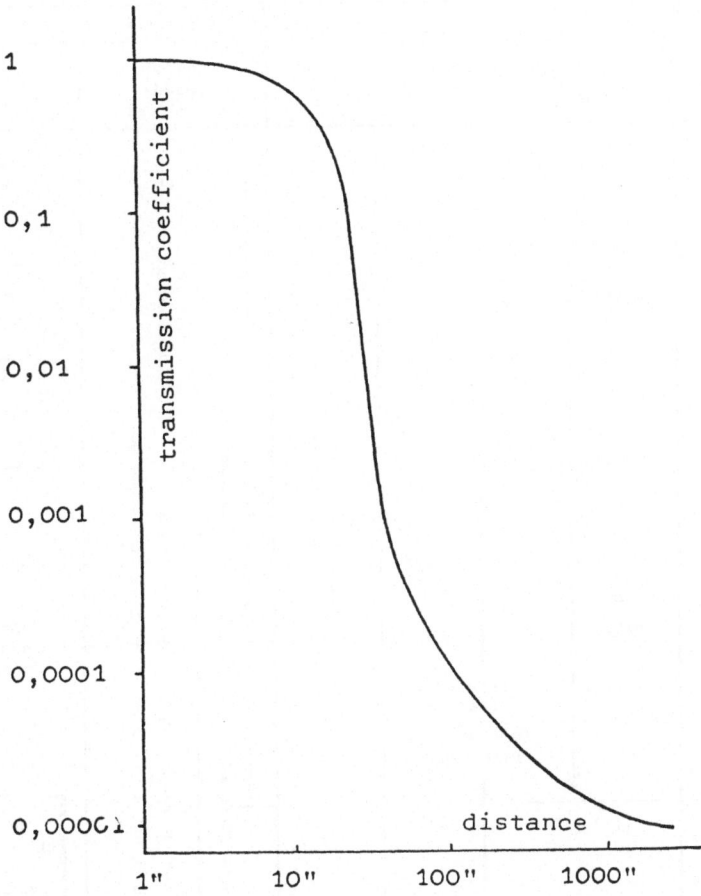

Fig. 1.   IFOV transmission profile. Transmission coefficient at any point inside the IFOV, as a function
of its distance to the center.

shown. Lindegren (1982) has given a more detailed such limit as a function of the
difference in magnitude Δm, taking into account *all scans* of the star inside the 2.5 years
observation period and *admitting that the star has no sensible orbital motion*. This leads
to the following table:

TABLE I

Primary magnitude B_1	ρ_{min} (arc sec)					
	$\Delta m = 0$	1	2	3	4	5
7	0.04	0.05	0.06	0.10	0.16	0.3
10	0.05	0.06	0.08	0.12	0.2	0.4
12	0.06	0.07	0.10	0.15	0.3	0.5
13	0.08	0.09	0.12	0.2	0.4	–

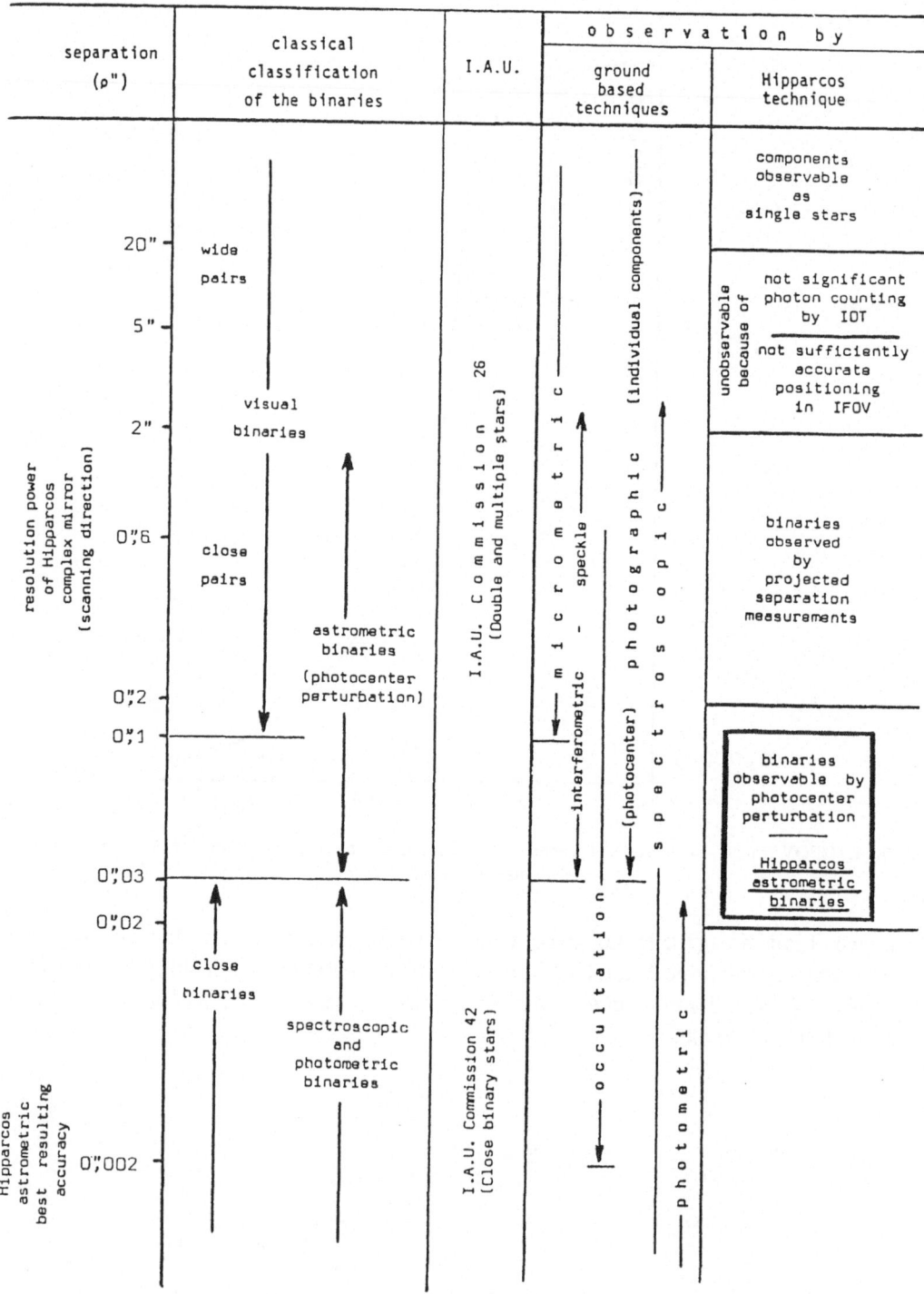

Fig. 2.

One can see that, in some favourable cases, the above-mentioned limit is shifted to $0''.04$. But for such a small separation of the components, only a systematic search for new pairs appears feasible, any accurate measurements of relative positions being perhaps not always impossible, but at least very difficult. For such measurements, it seems to be necessary to consider only distances larger than $0''.2$. Bacchus (1982) has also recalled that the recognition of the binary nature of a stellar image, may be easily done in many cases only on the basis of *one favourable scan* by considering the degree of coherence between the amplitudes of the terms of the different orders of the photometric signal analysis.

The limit of $0''.1$ is the separation of the components of the closest pairs that have been observed with ground based equipment by visual techniques (micrometric measurements).

(e) $\rho = 0''.03/0''.02$ is the rough boundary between astrometric binaries and pure spectroscopic ones. The smallest orbital perturbation effect that can be detected on the photocenter of a binary photographic image is of the order of $0''.01$, which corresponds to a semi-axis major greater than $0''.03$ (e.g., the inferior limit of the astrometric binary field, see Section 3). This boundary fits also more or less with the overlap region of both IAU Commissions Nos. 26 and 42 complementary activities. Its interest has nothing to do with any eventual 'commission's territorial claims' – their activities being clearly defined – but only with the different meanings that members of both commissions give to the expressions: 'close pairs' and 'close binaries'.

It is also the inferior limit for observation of binaries by interferometric and speckle techniques (McAlister and Hartkopf, 1984).

(f) $\rho = 0''.002$ is, finally, the Hipparcos astrometric accuracy and thus the distance between the components of a system under which no more discussion may take place here. It is also the extreme limit that the photometric occultation technique has reached in resolving close pairs (Evans, 1983).

It should be added that the diagram of Figure 2, has been established without generally taking into account that nearly all the limits between the different categories of binaries, are functions of the difference in magnitude between their components. All these limits are thus more or less blurred out and the diagram should be considered as a rough scheme to make easier any discussion on binary problems in the Hipparcos mission.

This especially will help in the case of the *astrometric binaries*, that we shall consider now.

3. The Astrometric Binaries by Ground Based Techniques and by the Hipparcos Technique

(a) The main difference between *astrometric binaries* and all other categories of double stars, lies in the fundamental role of their photographic photocenters. Discovery, systematic measurements and orbital computations of astrometric binaries are all based on the position of this particular point of the photographic images of the systems. This also implies the use of reference stars and the imbrication in a same problem, of orbit, parallaxe and proper motion determination.

If A is the center of the principal component; B, that of the secondary; G, the center of mass of the system; and P, the photocenter of its image, one has (Dommanget, 1953, 1969):

$$PG = AG - AP$$

and thus for the semi-axis major α'' of the photocentric orbit and for the semi-axis major a'' of the relative orbit of B around A:

$$\alpha'' = (k - \beta)a'' ,\tag{1}$$

k being the mass-ratio $\mathfrak{M}_B/\mathfrak{M}_{AB}$ and

$$\beta = \frac{1}{1 + 10^{0.4\Delta m}} ,\tag{2}$$

as proposed by Van de Kamp (1937) who has been with his staff at the Sproul Observatory (Swarthmore) the most involved astronomer in astrometric binary research.

The best accuracy for relative photographic measurements being of the order of somewhat better than $0''.01$ and the difference $(k - \beta)$ having a maximum of approximately 0.25 (for Δm of the order of $3m$) it appears that the astrometric technique may reach only systems having a component separation greater than $0''.03/0''.04$.

One also knows that for practical reasons, in case of ground based astrometric techniques, only binaries having a period between about 1 yr to approximately 25 to 50 yr, may be discovered. On one side, periods smaller than 1 yr are difficult to sort out

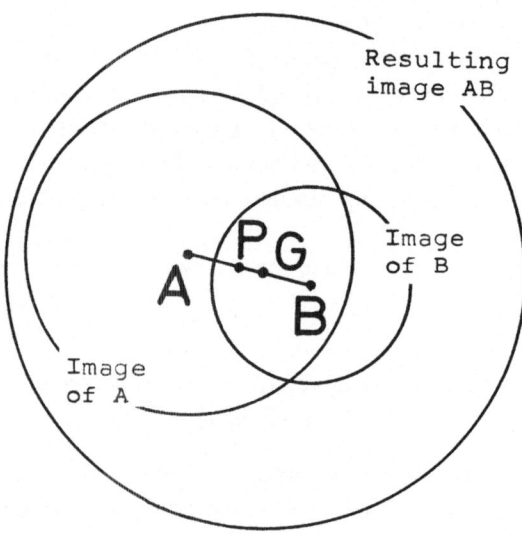

Fig. 3.

from a nearly one year periodical observation campain and also to separate from the one year parallactic motion; on the other hand, too large periods lead to a nearly linear orbital motion in the relatively short interval of observation time, that will be combined

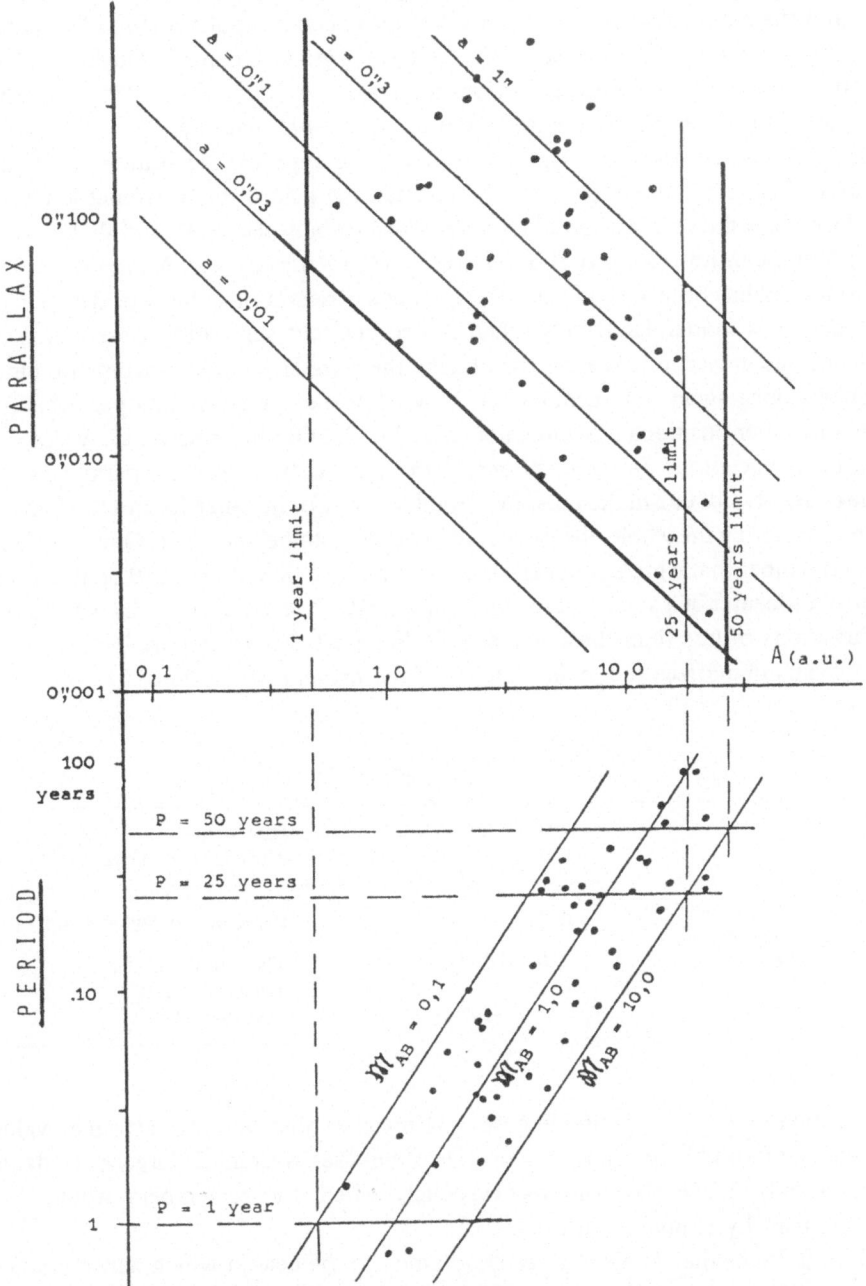

Fig. 4. Distribution of the presently known astrometric binaries, following their period, their parallaxes, their apparent semi-axis major (not photocentric) and their masses.

with the proper motion of the system, as we recalled it earlier (Dommanget, 1953). In Figure 4, the presently known astrometric binaries have been plotted in two bounded diagrammes ($\log \pi$, $\log A$) and ($\log P$, $\log A$) the units being for P, the year and for A, the astronomical unit. One can see the perfect agreement between the theoretical point of view and the practical situation, if one takes into account that because of formula (1), the semi-axis major a'' is at least always four times greater than α''. The limits of 1 and 25–50 yr appear clearly. The last one increased slowly from 25 to 50 yr in nearly half a century with the regular increase of the observational material.

One also can see from this figure that finally the apparent separation will generally not exceed slightly more than $1''$ to $2''$ and that the field of the astrometric technique for close pairs finally corresponds to *perturbations* between $0''.01$ and $0''.3$.

(b) What can now be said about *'possible' Hipparcos astrometric binaries*?

From what has been seen in case of Hipparcos observations, the consideration of the photocenter is fundamental only when the component separation is less than a few tenths of one arcsecond. We are then facing the *pure* Hipparcos astrometric binaries. The use – once again – of formula (1) shows that the corresponding perturbations of the photocenter positions may thus not exceed $0''.05$. On the other hand, we know that the expected accuracy on the positions, the proper motions and the parallaxes will be of the order of $\pm 0''.002$ in the most favourable cases (for instance for not too faint stars) when considering the whole observational material obtained in 2.5 yr. One consequently must remember that a greater uncertainty of some $\pm 0''.005$ to $\pm 0''.010$ on the positions of the photocenter of a stellar image *for each individual scan* must be considered and that perturbations of less than these values will thus probably be impossible to detect.

Table II summarizes the main features of the present discussion.

TABLE II

	Ground-based astrometric binaries	Hipparcos astrometric binaries
Photocenter accuracy	$\pm 0''.01$	Between $\pm 0''.005$ and $\pm 0''.010$
α''	$0''.01$ to $0''.3$	$0''.005$ to $0''.05$
a''	$0''.03$ to $1''-2''$	$0''.020$ to $0''.2-0''.3$
P	1 to 25–50 yr	Less than 10 yr

The 10-yr limit for the period in case of Hipparcos observations, is a rough value but it appears impossible to expect discoveries of binaries with much larger periods on the basis of only 2.5 yr observations, especially when one remembers what a binary measurement by Hipparcos means.

Thus, it seems that Table II gives good limits to the field in which new pairs may be *discovered* by photocentric perturbations during the Hipparcos mission. Compared to the field of possible discovery by ground based technique, the Hipparcos astrometric field is somewhat more extended to the smallest separations, but it is much more limited to the highest revolution periods.

Finally, the main interest of the Hipparcos technique will lie in the systematic character of the exploration of all stars brighter than 7 to 8 visual magnitude (the survey) and of many other fainter stars. *Discovery of new pairs should thus not be negligible.*

But the remaining question is: may such discoveries and observations lead to orbit computations and to new mass determinations?

4. Measurements and Computation of the Relative Motion in a Binary System Observed Only by Hipparcos

As one knows, Hipparcos has not been conceived for classical binary measurements. But this does not mean that it cannot produce such information. The main feature of an Hipparcos observation of a binary lies in the fact that it may only give for each scan – and when all other conditions permit it – the projection of the apparent distance between the components, in a known but variable direction: the scanning direction. The position angle of the system is thus unknown. If this is now variable for orbital reasons, the problem consists of removing the relative motion from the individual observations in order to permit the determination of the position, the parallax and the proper motion of the system with the highest accuracy.

In case where the orbit is known by other means (visual observations for instance), one has only to take this orbit into consideration and eventually, to adjust its elements by using the Hipparcos measurements.

In case where no orbit is known, it generally appears possible to determine the relative motion (orbital or linear) in the interval of 2.5 yr with sufficient accuracy as shown hereunder.

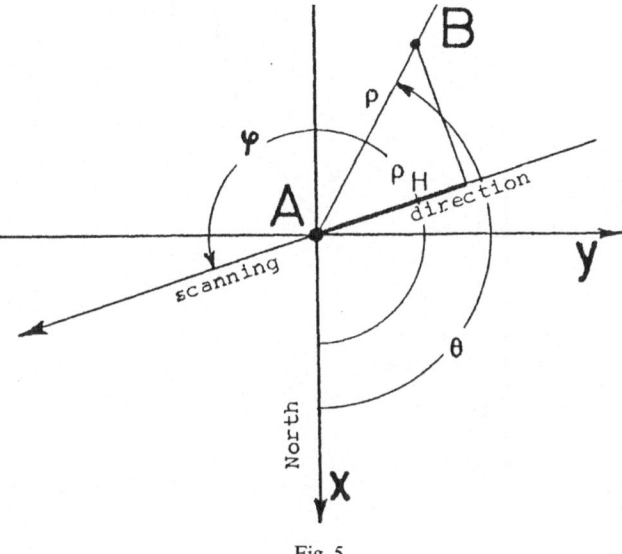

Fig. 5.

If A and B are both components, ϑ and ρ their position angle and their separation in a classical referential system Axy, and φ, the scanning angle in the same system (Figure 5), the corresponding Hipparcos measurement ρ_H is given by

$$\rho_H = \rho \cos |\vartheta - \varphi| \qquad (3)$$

and a complete measurement, by

$$(t, \varphi, \rho_H),$$

that one should usefully compare to a micrometric measurement which is given by

$$(t, \vartheta, \rho).$$

One should notice that ρ is essentially positive but that ρ_H is positive only when the scan reaches star A (primary) before star B (companion) and negative in the other case.

To make the discussion clearer, we shall consider a typical fictitious example where both components are observable (not an astrometric pair). The elements are:

$$P = 30 \text{ yr}, \qquad i = 40°00, \qquad m_{AB} = 7.7\,(\Delta m = 0.5),$$

$$a = 0''60, \qquad \Omega = 0.00, \qquad \mathcal{M}_{AB} = 1.16\,(\text{K1} - \text{K1}),$$

$$e = 0.40, \qquad \omega = 60°00, \qquad \pi = 0''059.$$

In such a case, one should first notice that the orbital motion in 2.5 yr, from 1988 to

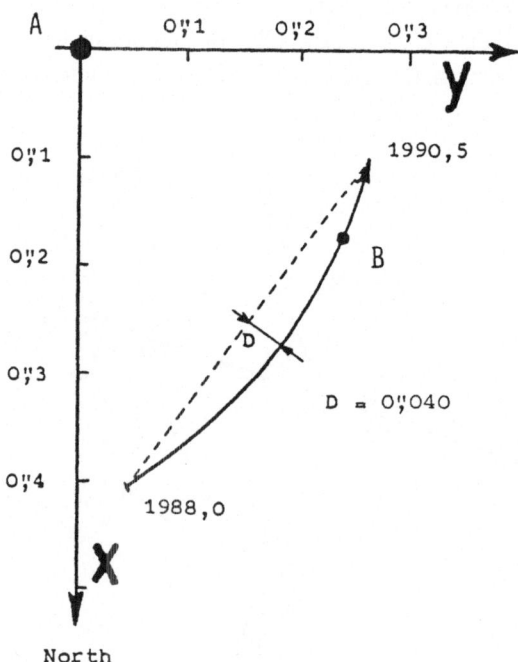

Fig. 6.

1990.5, if T = 1990, may not at all be approximated by a linear relative motion (Figure 6, maximum deviation between chord and orbit = 0″040!) and that it should thus be taken into account for position, proper motion and parallax reduction in the Hipparcos system.

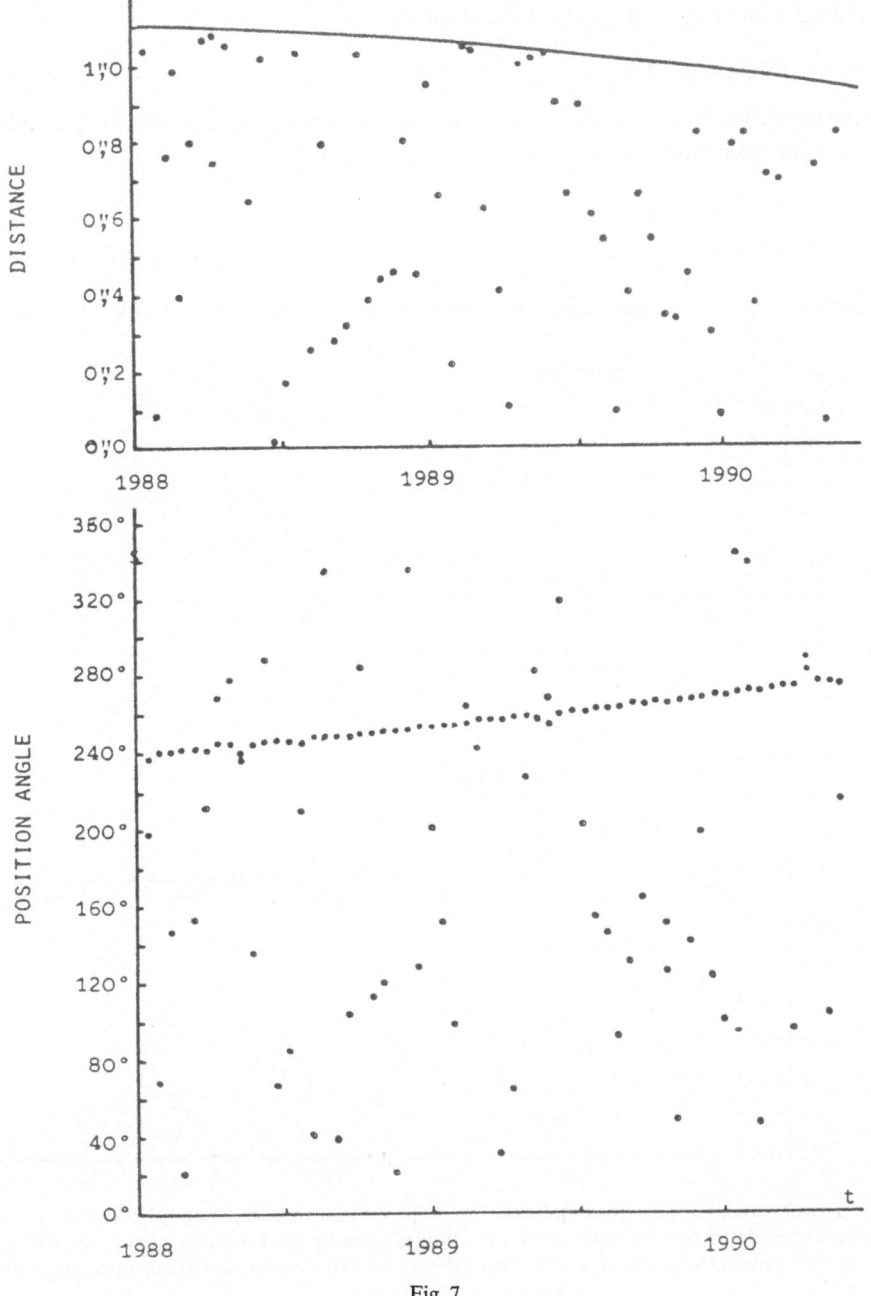

Fig. 7.

Suppose now first of all that the scanning directions appear at random between $0°$ and $360°$ in the 2.5 yr observation interval. The diagramme giving ρ_H as a function of the time t, will appear as shown on Figure 7. It is evident that the envelope-curve of the representative points fits fairly well with the function $\rho(t)$.

This envelope-curve being traced, one may read the corresponding ρ-values on it and compute the angle $\Delta = |\vartheta - \varphi|$ by the formula:

$$\cos\Delta = \rho_H/\rho \tag{4}$$

with the condition that $0° \leq \Delta < 180°$. For each scanning angle φ, one finds finally the only two possible values of ϑ:

$$\begin{cases} \vartheta_1 = \varphi - \Delta\,, \\ \vartheta_2 = \varphi + \Delta\,; \end{cases} \tag{5}$$

of which *only one* is acceptable. By plotting both values in a $\vartheta(t)$ diagram, the only

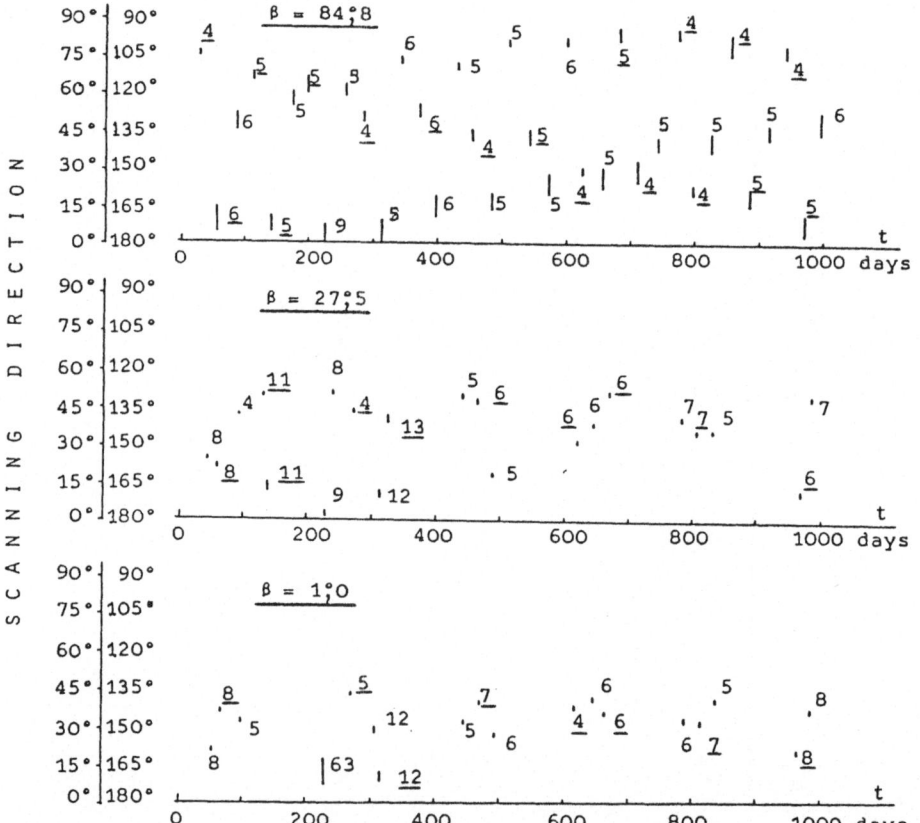

Fig. 8.   Distribution of the scanning directions by Hipparcos, for three different ecliptic latitudes. These directions appear by groups covering various amplitudes given by the extension of each segment and of which number is indicated nearby. The underlined figures refer to the second and fourth quadrants; the other to the first and third ones.

acceptable values of ϑ will immediately appear as belonging to a sequence of perfectly well distributed points on a curve which is the one we are looking for (Figure 7).

But, the Hipparcos scanning directions are not at random as can be seen in Figure 8 giving the distribution of the values of φ as a function of time, for three different typical star positions*. This makes the drawing of the envelope-curve somewhat more uncertain.

In case of the presently considered fictitious orbit, the first set of scanning directions has been used and a measuring error having $\sigma = \pm 0''.005$ has been adopted. Figure 9 gives the diagram of the corresponding simulated values of ρ_H as well as the values of ϑ_1 and of ϑ_2 deduced by (4) and (5), on the basis of the best envelope-curve drawn for the ρ_H-values. The diagram of ρ_H has been limited to $\rho_H > 0''.200$ knowing that all projected separations smaller than this limit are unmeasurable.

By the trial and error technique, the drawing of the final $\rho(t)$ curve is obtained for the best agreement between the corresponding points of the sequence observed on the $\vartheta(t)$ diagram. One should remark that the most accurate points in the $\vartheta(t)$ diagram are those corresponding to the smallest values of ρ, the Δ-values being then less affected by the errors of any kind on ρ. Simultaneously, the drawing of the curve $\vartheta(t)$ is thus also obtained. An orbit may then be computed as usually by any appropriate method. Using the Thiele–van den Bos method, we retained the following three fundamental points of which the agreement with the exact values given by the fictitious orbit is better than $1°.5$ in position angle and $0''.003$ in apparent separation.

	t	ϑ	ρ
I	1988.200	6°.000	0''.3980
II	1989.400	33.500	0.3300
III	1990.300	65.000	0.2860

On the basis of earlier researches conducted at Uccle and of which the practical aspects have been assembled in a general computer program by Nys (1983), we found, concerning the choice of the value of the double of the apparent areal constant that:

(a) a parabolic orbit appears for $C = +0.055\,487\ (C_p)$;

(b) elliptic orbits appear for C greater than C_p but smaller than $C = +0.055\,569$ (hyperbolic orbits appear for $C < C_p$).

Three elliptic orbits have been computed with values of C ($+0.055\,540;\ +0.055\,520;\ +0.055\,500$) inside its very short possible interval of variation. The periods found were respectively: 20, 40, and 163 yr but all the deduced ephemerides inside the 2.5 yr observation period are in very good agreement with the curves $\vartheta(t)$ and $\rho(t)$ read on the diagrams of Figure 9 and also with the relative motion computed with the elements of

---

* These scanning directions have been kindly communicated to us by M. Creze (Observatoire de Besançon) who is task-leader of the working group in charge of the simulation of the observation strategy, inside the Input Catalogue Consortium.

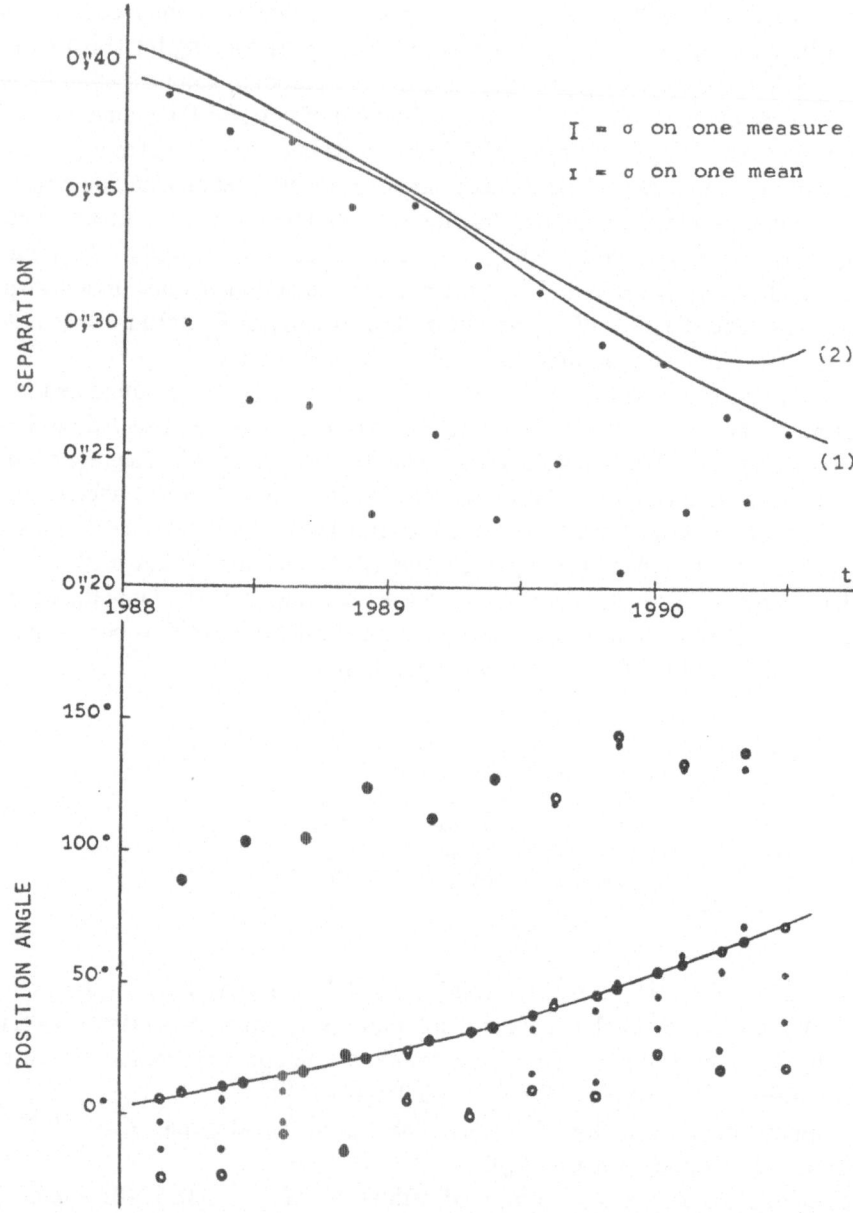

Fig. 9.   In the above diagram, each dot is the mean of the ρ_H-values given by the few successive scans performed at the same epoch. In position angle, the dots correspond to the envelope-curve (1) obtained by a first drawing; the circles correspond to the final adopted $\rho(t)$ curve (2).

the fictitious orbit. The computed values of ρ_H by any of the three orbits and those computed by the fictitious orbit, show differences less than $0\overset{\prime\prime}{.}016$ and a dispersion of $\sigma = \pm 0\overset{\prime\prime}{.}007$!

The use of the least-squares method seems not to lead, in the present case, to any

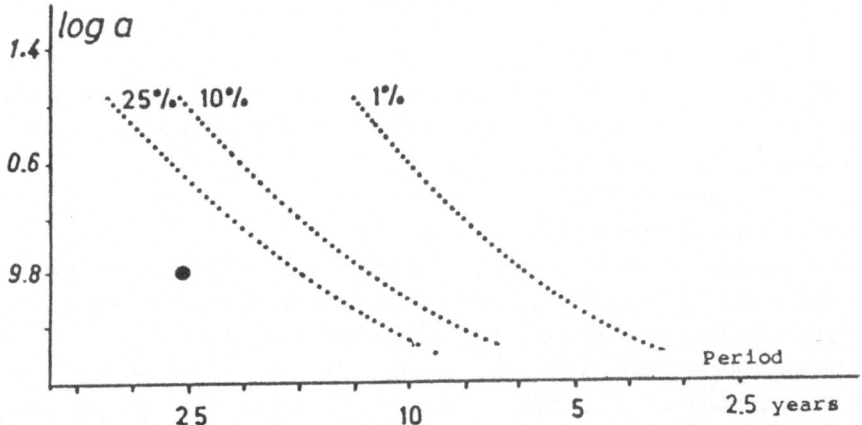

Fig. 10. Statistical error (in %) on the total masses deduced from orbits computed on the basis of observations regularly distributed inside a 2.5 yr observation interval and with an error of the order of $\pm 0''001$ (Dommanget, 1982a). The large dot represents the here considered fictitious orbit.

improvement of the solution. This confirms what we already have shown (Dommanget, 1982a) at the special session on 'Space Astrometry' held at Montreal in 1979 during the XVIIth General Assembly of the IAU about the *impossibility* of substantially increasing the quality of the presently known orbits of visual binaries by Hipparcos observations only and, as a consequence, neither the quality of the deduced total masses (Figure 10).

It thus appears that in the case where both components may be distinguished in the signal analysis and their relative position may be deduced (projected separation), their relative motion may be accurately determined inside the 2.5 yr observation interval. After having removed this motion from the observations, the parallax and the proper motion, as well as the position, may be computed with the same accuracy as for single stars. But *no reliable orbital elements may be obtained.*

Now, if only the photocenter may be observed (astrometric binaries), the situation is much more complicated, as it is in the case of astrometric binaries observed by ground based techniques. The fundamental problem remains the location of the center of mass of each system. Only in case of short periods (less than two or three years) one may hope for some solution: as soon as this is possible, the technique we have shown on the basis of our fictitious orbit may be used not only for relative motion determination but sometimes (because of the shortness of the period) for reliable orbital elements determination. The least squares method may be in that case of some help. But the situation will also mainly depend on the importance of the parallax of the system and of the possibility of separating both effects: parallactic and orbital motion.

Finally it appears in our opinion, that nearly each case will be a particular one and that it would be difficult to propose a general investigation technique that would be applicable to all systems.

5. Conclusions

Remembering that a double star with a component separation between $2''$ and $20''$ will generally not be observable by Hipparcos, three main different categories of double stars should thus be considered for this space mission, from the present discussion:

- the wide pairs with $20'' < \rho$,
- the close pairs with $0''.2 < \rho < 2''$,
- the astrometric pairs with $0''.02 < \rho < 0''.2$.

Concerning the *wide pairs*, parallaxes and proper motion deduced for both components will greatly help in any discussion on the eventual opticity of the pairs, by using appropriate criteria (Dommanget, 1955, 1956, 1960, 1966, 1967).

In the case of *close pairs*, the relative motion of their components may be computed. Only in exceptional cases, it will lead to the determination of a reliable orbit. The knowledge of the relative orbital motion from the Hipparcos measurements (eventually combined with ground-based measurements) will permit to obtain the best values for proper motion and parallax. Our knowledge on the masses will be improved by an increase of the amount of the material more than by any substantial increase of its quality. Discovery of new pairs will probably be another important result of the Hipparcos mission.

Finally in the case of *astrometric binaries*, more than in any other cases, orbit, parallax, proper motion and position determination will set a very difficult problem that probably will have to be solved differently following the various categories of situations. Therefore, in place of making further comments, it seems to be more realistic to 'wait and see' what really the observations of astrometric binaries by Hipparcos are going to be.

But what anyway will be of the highest interest is the identification by Hipparcos of all stars in the *Input Catalogue* that will show photocentric perturbations larger than say $0''.005$. Even without any other information, it will be of an extra-ordinary help for instance in organizing new campains of research by ground-based observations.

The Hipparcos equipment should thus be considered as an important tool of research in double star astronomy and especially in case of astrometric pairs of which Hipparcos will particularly improve the present census.

Acknowledgement

We wish to express our sincere thanks to our colleagues O. Nys and J.-L. Overal, for their contribution to the computations and work needed by the preparation of the present paper.

References

Bacchus, P.: 1982, *Caractéristiques des étoiles doubles observables par Hipparcos*, International Colloquium on the Scientific Aspects of The Hipparcos Mission, Strasbourg 1982, ESA SP-177, pp. 153–159.
Bacchus, P.: 1984, *Observabilité des étoiles doubles par Hipparcos*, Inca News, No. 5, pp. 20–26.
Dommanget, J.: 1953, *Ciel et Terre* **59**, 229.

Dommanget, J.: 1955, *Bull. Astron.* **20**, 1.

Dommanget, J.: 1956, *Bull. Astron.* **20**, 183.

Dommanget, J.: 1960, *Bull. Astron.* **23**, 101.

Dommanget, J.: 1966, *Bull. Astron.* **29**, 511.

Dommanget, J.: 1967, 'On the Evolution of Double Stars', *Colloque UAI*, Uccle, 1966, p. 25.

Dommanget, J.: 1969, *Ciel et Terre* **85**, 75.

Dommanget, J.: 1982a, *Astronomie* **96**, 15.

Dommanget, J.: 1982b, *The Interest of Double-Star Observations by Hipparcos*, International Colloquium on the Scientific Aspects of The Hipparcos Mission, Strasbourg 1982, ESA SP-177, pp. 161–164.

Dommanget, J.: 1983a, 'L'Astronomie des étoiles doubles face à la mission astrométrique du satellite Hipparcos', *Colloque UAI* **62**, Comm. Obs. R. de Belg. B, No. 127.

Dommanget, J.: 1983b, *Bull. Inform. Centre Données Stellaires* **24**, 83.

Evans, D. S.: 1983, 'Catalogue of Occultation Double Star Observations, *Colloque UAI* **62**, p. 73.

Lindegren, L.: 1982, *Imaging Properties of Hipparcos and the Observation of Multiple Stars*, International Colloquium on the Scientific Aspects of The Hipparcos Mission, Strasbourg 1982, ESA SP-177, pp. 147–152.

McAlister, H. A. and Hartkopf, W. I.: 1984, *Catalog of Interferometric Measurements of Binary Stars*, Chara Contribution, No. 1.

Nys, O.: 1983, 'Programmes for Computing Visual Binary Orbits from Three Fundamental Points and the Apparent Areal Constant', *Colloque UAI* **62**, Comm. Obs. R. de Belg. B, No. 128.

Van de Kamp, P.: 1937, *Astron. J.* **46**, 36.

HIGH RESOLUTION PROBLEMS IN ASTROMETRY AND SPECTROSCOPY OF BINARIES AND THEIR SIGNIFICANCE FOR THE PROGRESS OF STELLAR PHYSICS*

FRANS VAN 'T VEER

Institut d'Astrophysique, Paris, France

(Received 9 August, 1984)

Abstract. Two centuries were necessary for binaries to become, after having been unknown and ignored, one of the outstanding subjects of stellar research. This remarkable evolution has resulted now in the reversed situation with the painful question, at least for some of us: do single stars really exist?

It is recognized by everybody, that many essential properties of stars can only be obtained from observations of close and wide binaries.

It is the aim of this contribution to see how binaries can help us to better understand binaries and single stars, even if the existence of the latter is not yet definitely confirmed.

1. The Anticipation of Bessel

F. W. Bessel was a great friend of W. Olbers, the physician-amateur astronomer who discovered Pallas, Vesta, and six comets. Their long and attractive correspondence tells us better than any paper about the convictions and motivations, disappointments and joys that Bessel experienced in his work and life.

From Bessel's letter of 7 March, 1812 we can learn about his fertile intuition. He describes there the binary 61 Cygni and his determination of the high proper motion of its two components. He concluded that the system must be a physical binary and not an apparent one as was still believed by many others at that time. In the same letter he suggests, considering the high proper motion, that 61 Cyg should have a large and measurable parallax. Both conclusions were true; 61 Cyg is a physical double star and 26 years later Bessel was able to correctly determine its parallax, as can be read in the paper given by E. H. Geyer at this conference. It was the first parallax ever measured. In what follows parallaxes of binary stars will play an important role.

The aim of my talk this afternoon is to discuss the question what stellar astrophysicists can learn from the study of all sorts of close and wide physical binaries: visual and astrometric binaries, photometric and spectroscopic binaries. I shall limit myself in this short review to the problems related to solar mass stars. The reason for this is easy to understand. The observed massive stars are always young with respect to the age of the Galaxy. The much more slowly evolving solar type stars on the contrary can belong to regions where star formation is still going on as well as to old clusters. Their internal structure is directly related to the chemical composition of the galactic disc at different

---

* Communication presented at the International Conference on 'Astrometric Binaries', held on 13–15 June, 1984, at the Remeis-Sternwarte Bamberg, Germany, to commemorate the 200th anniversary of the birth of Friedrich Wilhelm Bessel (1784–1846).

epochs and different places and, hence, I think that the less massive stars are of great interest when we wish to understand the evolution of the Galaxy.

2. The Rapid Development of Instrumental Techniques not only solves our Problems

The astronomers of the 19th century inaugurated three different important research fields related to double star astronomy. The determination of the relative orbits of binaries at the beginning, the measurements of the parallaxes in the middle and the spectroscopic analysis of radial velocities towards the end of the century. All these three fields have been developed since that time, almost without interruption. As a matter of fact, the mean error of individual measurements of parallaxes, radial velocities and orbital diameters has decreased by a factor of 10 about every 50 years. Near the end of the present century we shall be able to measure milliarc seconds (mas) and velocities in the 10 m s^{-1} range.

The big questions I want to discuss with you at this meeting are the following:

What can astrophysicists do with all the material expected from new observational techniques?

What is the precision they need to make progress in astrophysics?

What do we understand by progress in astrophysics?

We do not want to satisfy ourselves with the easy answer that new positions, distances, and orbits and more precise color-magnitude diagrams are a sufficient scientific aim in themselves. The new data will surely give us a better knowledge of the population of the near solar neighbourhood. However, according to Hipparcos parallaxes with an estimated mean error (m.e.) of $0''002$ we shall obtain for distances of 50 pc stellar positions from individual measurements with a m.e. not better than 10%. Hence, the error of absolute magnitudes will be $\Delta M_V = 0.2$ which is insufficient for an interesting interpretation of the $C - M$ diagram. Good interpretations of the $C - M$ diagram require accuracies better than $\Delta M_V = 0.1$. This means that the stars must be nearer than 25 pc.

Now I shall immediately ask the question: What is the significance of the $C - M$ diagram for the progress of modern astrophysical research? The answer must be realistic and not influenced by emotional attachments to pioneers as Russell, Hertzsprung, and so many others after them.

We must not forget that the $C - M$ diagram is essentially a classification scheme, developed after the early trials of Secchi and others, to order stars of different colors and luminosities. Only later it could be developped into a new and powerful tool for determining the evolutionary stage of stellar groups supposed to be of the same age, and even to recognize stars which are different from the other members of the group. It has always been, and is still now, our lack of imaginative insight which makes us often think that nature repeats itself identically according to some divine blueprints set up once and forever. In the early days of the invention of the two-dimensional $C - M$ diagram it was believed by many people that two stars which occupy the same position therein would be physically identical. From trials to construct 3- and even 4-dimensional diagrams it became more and more clear that the different sorts of stars, that nature can make out of initially slightly different constituents in differing environments, cannot be described by a limited number of parameters. All stars are different, and when we see two which seem to be identical, we may be sure that our instruments are still unable to detect the differences. However, there are at least 10^{11} stars in our Galaxy and our aim cannot be to analyze them all even when they are different. Moreover, the remarks made above about stars are also valid for clouds, meteorites, quasars, galaxies...

3. What Do We Want to Know about Stars?

We urgently need an answer to the following question: What do we want to know about stars? and connect with it: What is our task as astronomers and astrophysicists?

For Bessel the answer to these two questions was simple and direct (Bessel, 1841): "Was die Astronomie leisten muss, ist zu alle Zeiten gleich klar gewesen, sie muss Vorschriften erteilen, nach welchen die Bewegungen der Himmelskörper, so wie sie uns von der Erde aus erscheinen, berechnet werden können. Alles, was man sonst noch von den Himmelskörpern erfahren kann, z.B. ihr Aussehen und die Beschaffenheit ihrer Oberflächen ist zwar der Aufmerksamkeit nicht unwert, allein das eigentliche astronomische Interesse berührt es nicht... die Bewegungen aller Himmelskörper so vollständig kennen zu lernen, dass für jede Zeit genügend Rechenschaft gegeben werden kann das war und ist die Aufgabe, welche die Astronomie zu lösen hat."

Astrophysics as a discipline of research did not exist at that time, and the spatial motions and positions of the celestial bodies were the only things that could be studied quantitatively. A more recent answer to our question, showing the influence of more than a century of physics and philosophical thinking can be found in a critical study of Popper (1974): "The task of science is not to collect data, but to find satisfactory explanations." We now want to know what is meant by satisfactory explanations?

Popper's answer may help us: "An explanation in terms of testable universal laws and initial conditions is satisfactory."

We know that our astrophysical theories are of a transitory nature, but philosophical and logical statements are temporal as well. For example, few people still adhere to the determinism according to Laplace who considered the Universe as a big clockwork mechanism. The working principle of Popper is perhaps still too mechanistic for some of us, and is surely not free from hypotheses. However, it sounds sane to try to formulate initial conditions and general laws involving the present temporal and continuously changing situation as it is suggested by Popper. It is just this presently observable situation which we can study by collecting the new observational data.

4. How Can We Learn More about Stars?

In the next sections I shall try to examine, following the ideas given in the preceding section, a limited number of problems that better binary orbits and parallaxes, and more precise radial velocity curves may help us to solve in a predictable near future. In the same context I shall also consider the light curves of eclipsing binaries obtained from ground-based observatories. For these pure photometric data we may expect that the precision, limited by the Earth's atmosphere, will remain about the same.

The subject is concerned with topics of direct astrophysical importance such as the study of the density, the internal chemical composition and spectrophotometric properties of the stars in general. All these topics are related by the common property that binaries have given a substantial contribution to their understanding and will continue to do so for still a long time. At every possible occasion, observed parameters will be confronted with those computed from stellar evolutionary models. These models, of which many examples can be found in the literature, are based on a limited number of initial conditions, auxiliary parameters and hypotheses concerning the physical processes. From the confrontation of this model with parameters derived from observational data, one may hope to obtain better physical explanations and greater insight into the initial situation. If this is no longer possible, one or more branches of the study of binaries will have to be revised or abandoned.

5. Stellar Parameters from Close Binaries

We shall first discuss what we can learn about the density of the stellar material. From close binary stars we can gain direct information, at least in some favorable cases, about the density structure of the stellar material. The internal density concentration may be studied using the apsidal motion in eclipsing binaries (see, for example, Kopal, 1940; or Giménez and Garcia-Pelayo, 1982) and used as a test for the models and, hence, our understanding of the stars.

The mean density $\langle \rho \rangle$ of a star can evidently be determined from its total mass and radius. It is an essential mass-dependent parameter among the zero-age characteristics and, furthermore, directly dependent, for a given mass, on the age of the star.

The only way of determining stellar masses, without invoking hypotheses about stellar physics, stellar constitution, or both, is via the combined determination of $M_1 + M_2$ (from the period P and the semi-major axis a) and $q = M_2/M_1$ (from radial velocity curves of both components of a binary or the astrometric determination of the center of gravity of a visual binary).

The diagram of Figure 1 shows us the various ways and combinations which are possible for obtaining stellar masses. The highest precision can be obtained, for the time being, from components of eclipsing binaries which show a resolvable double-lined spectrum. In exceptional cases a precision of a few percent may be obtained (Andersen *et al.*, 1980) but, generally, we have to content ourselves with a mean error greater than 10% (see Popper, 1980, for an interesting account about the question of precision).

The best way, up to now, to determine stellar radii of non-giant components is, again, the combined study of eclipsing and spectroscopic binaries. Mean errors of a few percent only can be obtained in favorable cases.

From observed masses (m), radii (r) parallaxes, and colors (C) we may determine relations between the luminosity (L) and m, r and m, L and r and so on. This work

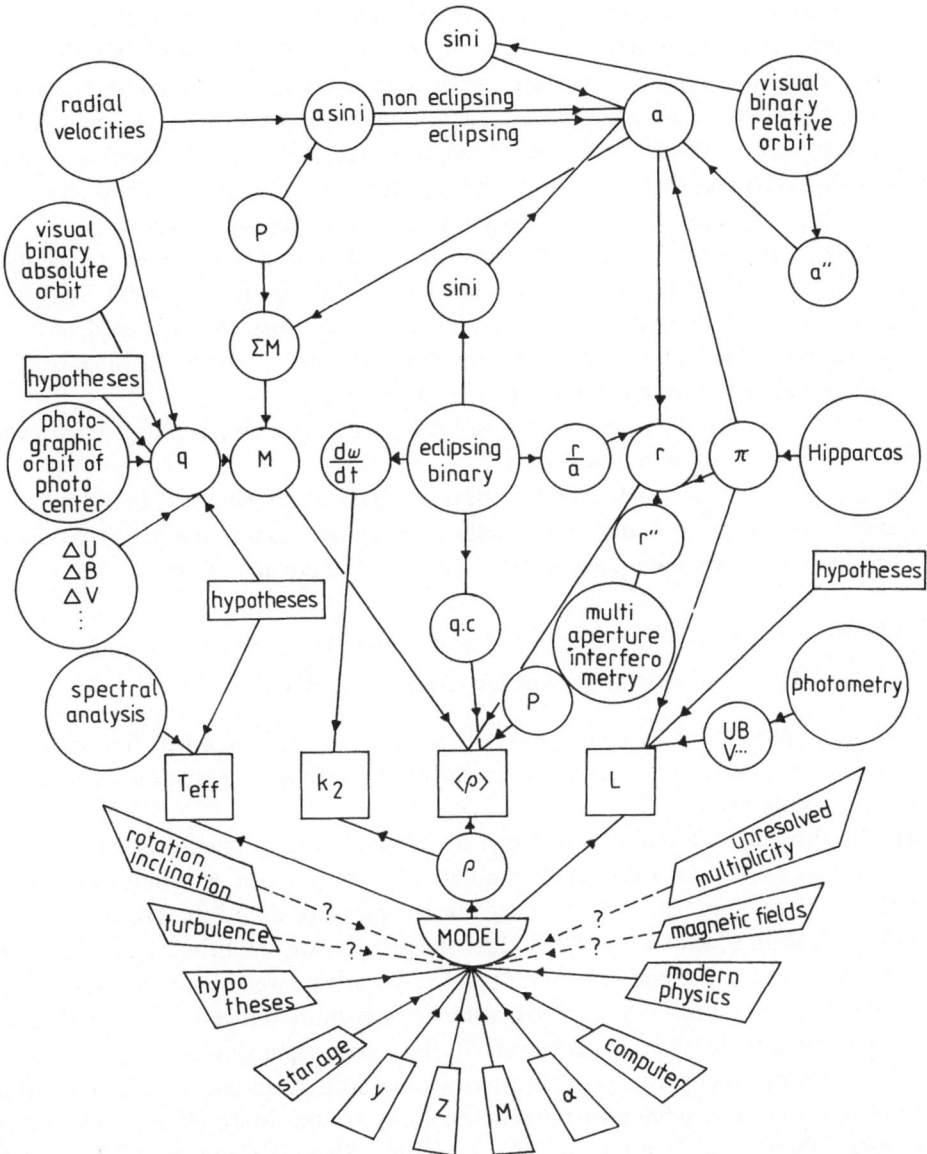

Fig. 1. This diagram shows the different ways to confront observational analyses with theoretical syntheses. At least until the end of this century eclipsing binaries will continue to give the best astrophysical information. Great circles indicate the source, that is to say the sort of work and/or the sort of binary where the information is coming from. This information is indicated in small circles by arrows which relate it with the source, or different sources when two or three arrows meet each other in the same point on the small circle. Symbols: a, true orbital major semi-axis; a'', apparent orbital major semi-axis; c, degree of contact for contact binaries; i, orbital inclination; k_2, coefficient of density concentration; L, absolute total luminosity; M, mass; P, period; π, parallax; q, mass ratio of the components; r, true radius; r'', apparent radius; ρ, density function; T_{eff}, effective temperature; U, B, V, absolute magnitudes; ΔU, ΔB, ΔV, magnitude differences between components; ω, longitude of periastron. L, M, r, r'', ρ, T_{eff}, U, B, V, are in relation to one of the components of the system. The chemical composition (Y, Z) and the mixing length-pressure scale-height ratio (α) are three essential parameters for the construction of evolutionary stellar models.

has been done several times (see especially Habets and Heintze, 1981) and it is far from sure that further important improvements can still be made in the same way. We must not forget that all these relations depend at least on the age and the chemical composition of the stars. As long as these parameters cannot be properly taken into account, it has little sense to improve the existing 'mean' relation. It is certainly more interesting to study special individual cases than to 'improve' a mean curve obtained from individual values which strongly depend on badly known parameters (i.e., the stellar age).

I entirely share the opinion of Andersen *et al.* (1980) that it is senseless to calculate absolute dimensions with a precision of some tens of percent. If we not only intend to put m and r in a catalogue of stellar parameters, but also wish to use their diagnostic values to improve the input physics and chemistry of stellar evolution (see Figure 1), it is advisable to strive after no more than 1% mean errors.

Are there other ways than the analysis of photometric and spectroscopic binaries to do precise work on absolute dimensions?

For example, what can we do with the observations of visual binaries? In the diagram of Figure 1, where arrows indicate the method mentioned above, alternative possibilities, to find the intermediate parameters a, $\sin i$, q, etc., are also given.

6. Stellar Parameters from Wide Binaries

I shall now analyze the different possibilities that may be combined for obtaining intermediate parameters in other ways than those derived from photometric and spectroscopic close binaries.

(1) The absolute semi-major axis, a, can be obtained from its apparent value and the parallax. If we want a final precision of about 1% for $\langle \rho \rangle$ it is absolutely necessary that about the same precision is obtained for the parallax π and the apparent major semi-axis a''. With a mean error of $0''.002$ for the parallax we are, hence, limited to stars nearer than $0''.2$ or $0''.1$ when several measurements can be weighted to a mean value. The quality of orbital solutions for visual binaries is more difficult to judge. Unfortunately, it is impossible to find mean errors for the orbital separation of binary solutions given in papers or catalogues. I estimate, with a good dose of optimism that it is possible to obtain for some tens of binaries, the semi-major axis of the orbital ellipse, with a mean error better than 1% before the end of the century. Since the total mass of a binary depends on the third power of the orbital radius, this extreme effort implies nevertheless a 3% error for the total mass of the system.

(2) The mass ratio q can be obtained, without the use of a spectrograph, from the motion of one of the components with respect to an external reference frame. It may be hoped that in a near future some good mass ratios may be determined for close visual binaries. On the other hand some mass ratios can be determined from astrometric photographs of photographically unresolved visual binaries. The results, however, depend on the transformation from photocenter to barycenter, which in its turn depends on the magnitude difference of the two components. The precision of this sort of

measurements is difficult to appreciate, but mean errors better than 10% appear to be very rare (Feierman, 1971).

A third way, also indicated in the diagram, to determine the mass ratios may be tried from spectral analysis or simply the magnitude difference of the two components supposed to be Main-Sequence stars. Evidently, age and chemical composition play their role in the application of this method. So this indirect way of working has to be avoided, for it introduces implicitly parameters we just want to determine.

(3) The stellar radius may be determined from the light curve of eclipsing binaries when the distance is known. Another well-known, but quite different, method makes use of interferometric techniques. However, even with future resolutions of 10^{-4} mas, as may be expected before the end of the century, the 1% criterion can be reached only for stellar diameters greater than $0\overset{''}{.}01$ that is something like the solar radius at 1 pc. Hence, it is reasonable to expect that the 1% error level cannot be obtained in the near future for solar-type Main-Sequence stars by high-angular resolution interferometry.

(4) We arrive at the conclusion that the combination photometric-spectroscopic binary will still remain the best to be explored for the next decades to come, at least for the astrophysical needs as defined above.

Even the combination visual-spectroscopic binary cannot help us very much in near future, for no precise stellar radii can be found in that way and also the mass determination remains uncertain. The velocity in a circular Kepler orbit is inversely proportional to $a^{1/2}$ or $P^{1/3}$. When it will be possible to measure radial velocities, with a mean error of less than 1%, down to 1 km s^{-1} many new binaries will become accessible for a precise determination of the absolute dimensions.

However, there is one problem that needs to be considered with special attention. According to a study of Dravins (1975), the interpretation of stellar radial velocities smaller than 0.5 km s^{-1} must be done with caution. Small inhomogeneities of the stellar photospheres above a convective zone, will be at the origin of wavelength shifts and asymmetries of the spectral lines. These perturbing effects will make it difficult to directly relate the motion of the star to the measured mean displacement of the spectral lines measured by the high resolution multi-slot devices.

Hence, a new theory for the interpretation of measured mean velocities of the photocenter of the components has to be developed and applied. The interpretation of the data always becomes more complicated when the performances of the instrumental equipment increase.

7. Spectrophotometric Parameters

In the preceding paragraphs we examined the mean density $\langle \rho \rangle$ as a parameter with diagnostic value for theories of stellar astrophysics. As is illustrated schematically in the diagram of Figure 1, $\langle \rho \rangle$ may be compared with theoretical values obtained by model computations. At present, only binaries provide us with direct information about stellar densities, but still some other parameters are available for confrontation between observation and theory.

The diagram shows two spectrophotometric parameters, the effective temperature T_{eff} and L, the total luminosity or bolometric magnitude. Both parameters can be calculated and measured for individual stars and binary components as well, hence, they do not need duplicity as a necessary condition. The aim of the confrontation between observed and model-calculated parameters is a better understanding of the physics of the stars. Therefore, it is necessary that a star be entirely defined by a limited number of input parameters. The most important of them are given in the diagram. It is hoped that the influence of many second-order parameters like rotation (uniform or non-uniform), inclination of the rotation axis, turbulence, unresolved multiplicity, and so on, will affect the stellar constitution to a smaller degree.

Moreover, we must not forget that T_{eff} and M_{bol} cannot be measured directly. They have to be derived from measurements at different wavelengths which need to be combined. Therefore, a certain number of hypotheses concerning the structure of the stellar photosphere are implicitly present in the 'measured' values of T_{eff} and M_{bol}. In the next paragraph we shall try to see what can be learned from the comparison of calculated and observed properties.

8. From Parameters to Satisfactory Explanations

The underlying idea is simple, too simple, for it supposes that stars may be described by a limited number of parameters. By supposing this we are negligent as were astronomers of the beginning of this century, who thought that two parameters were sufficient to learn about the essential facts about the stars. Two stars are never identical, but they may have uncertain number of common characteristics. It is the knowledge of these characteristics which may help us to better understand what stars are alike. On the other hand, we can also learn, from the study of binaries, about the early development of the Universe. For example, the big bang and the resulting initial chemical composition can be studied by the determination of M_{bol}, T_{eff}, and $\langle \rho \rangle$ in the following way: the diagram of Figure 2 represents the (Y, Z, α)-space which may be used to record the parameters of the evolutionary model of a star of given age (t) and mass (m) which are fixed for the whole diagram. Every point represents there a well-defined model for which for example T_{eff}, M_{bol}, and $\langle \rho \rangle$ or k_2 the density concentration coefficient may be calculated.

When we now determine from observations three observable parameters for a star of known age (e.g., cluster age) and mass (from the binary data), it is possible (in principle) to determine directly, from the corresponding 3-dimensional diagram, Y, Z, and α. This means that we are able to determine the parameters of the initial chemical composition of the cloud out of which the star was formed and the mixing length-pressure scale-height ratio of the convective zone. All these data are of great importance for the understanding of the stars, the evolution of the Galaxy and the formation of the elements. It would also be interesting to find a lower limit for the Y (He-content) in very old stars, in connection with cosmological theories. It should also be possible to compare Y-values measured at the surfaces with those found from stellar models as

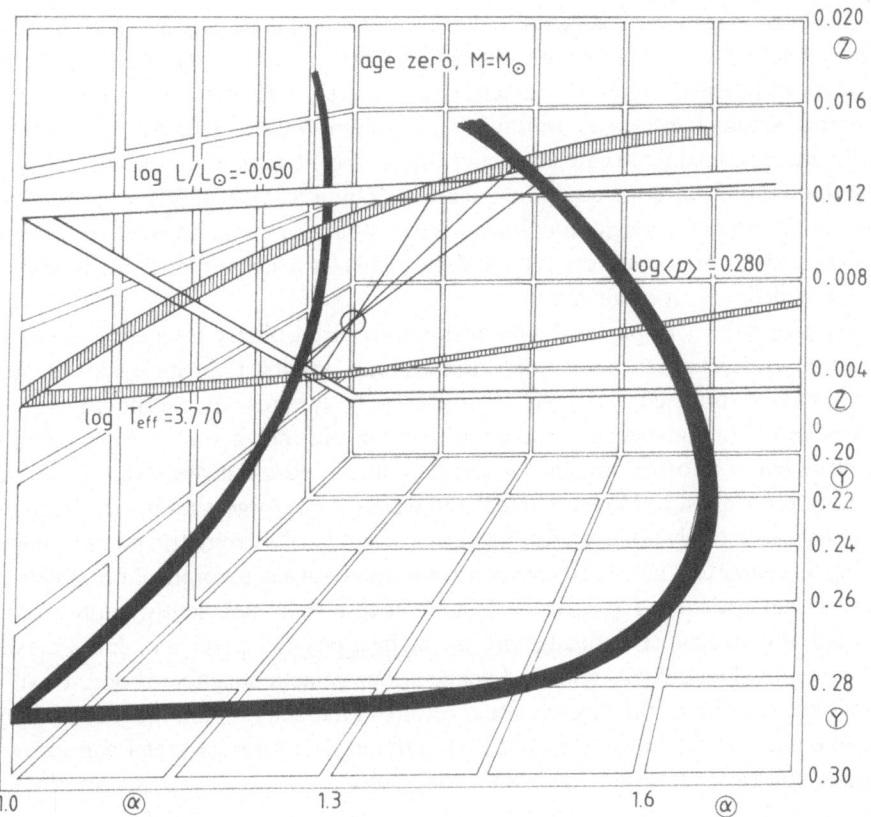

Fig. 2. For binary components of known age and mass it is, in principle, possible to determine Y, Z and α from $\langle \rho \rangle$, T_{eff} and L. The diagram illustrates the hypothetical case of a zero-age, one solar mass component with observed parameters: $\log \langle \rho \rangle$ (cgs) = 0.280; $\log T_{\text{eff}}$ = 3.770; $\log L/L_{\odot}$ = -0.050.

described above and new methods need to be developped to make such a comparison possible.

A special case is presented by contact binaries with known Roche surface structure. In that case the mass ratio q and the relative radius of the primary component are determined. With the aid of Kepler's third law, its mean density may be then be calculated and with estimated values of L, T_{eff}, and the mass, the age may be tested. I do not want to develop this point still further now, more information can be found in recent work on the four contact binaries of the old galactic cluster NGC 188 (van 't Veer, 1984).

9. Typical Binary Problems

Finally, I want to speak about some problems which are more specifically related to the binary nature of the stars.

In the beginning of the last century, double stars were gradually admitted as physical binaries with orbiting components. Since that time the new problem to be solved was no longer concerned with the existence of binaries, but with the question of their percentage among the stars, or groups of stars, of our Galaxy. This percentage has been steadily increasing and we can say, according to recent estimates (see, e.g., Gieseking, 1983) that virtually all stars could be members of binaries or multiple systems. Besides, those which are really single may have been double stars which lost their companion by mass transfer or encounters (van 't Veer, 1983). It is even impossible to claim that the Sun is definitely a single star.

Considering the astrophysical side of the problem of binary frequency, we face one of the most exciting questions of star formation. Is it possible that single stars are formed directly out of interstellar matter? Are planetary systems an alternative for binary systems necessitating special initial conditions, or are they a sort of endproduct from very close binaries losing angular momentum, as I tried to suggest (van 't Veer, 1975, 1979). All these questions are unsolved, and we must not forget that the most important questions are certainly not even formulated. The study of astrometric binaries may help to bring us nearer to a more decisive confrontation between theoretical possibilities and observational evidence. To reach this last aim it cannot be sufficiently emphasized how important it is to explore critically with the highest possible precision, and a maximum of observational techniques a limited space in our galactic neighbourhood. It will then be possible to combine the observational results in a much more homogeneous way. At the same time we shall discover, and that is the easiest forecast that I can make, that nature is still richer than we can try to guess now.

References

Andersen, J., Clausen, J. V., and Nordström, B.: 1980, in M. J. Plavec, D. M. Popper, and R. K. Ulrich (eds.), 'Close Binary Stars: Observations and Interpretation', *IAU Symp.* **88**, 81.

Bessel, F. W.: 1984, *Astronomische Untersuchungen I*, p. 32; quoted by D. Wattenberg, Archenhold Sternwarte Berlin-Treptow, Vorträge und Schriften, No. 2.

Dravins, D.: 1975, *Astron. Astrophys.* **43**, 45.

Feierman, B. H.: 1971, *Astron. J.* **76**, 73.

Gieseking, F.: 1983, 'Les étoiles binaires dans le diagramme H–R', *Comptes rendus sur les journées de Strasbourg*, t. 5, p. 4.

Giménez, A. and Garcia-Pelayo, J. M.: 1981, *Astrophys. Space Sci.* **92**, 203.

Habets, G. and Heintze, J.: 1981, *Astron. Astrophys. Suppl. Ser.* **46**, 193.

Kopal, Z.: 1940, *Harvard Obs. Circular*, No. 443, p. 1.

Popper, D. M.: 1980, *Ann. Rev. Astron. Astrophys.* **18**, 115.

Popper, K.R.: 1974, *Objective Knowledge*, Clarendon Press, Oxford, p. 203.

Van 't Veer, F.: 1975, *Astron. Astrophys.* **40**, 167.

Van 't Veer, F.: 1979, *Astron. Astrophys.* **80**, 287.

Van 't Veer, F.: 1983, 'Les étoiles binaires dans le diagramme H–R', *Comptes rendus sur les journées de Strasbourg*, t. 5, p. 145.

Van 't Veer, F.: 1984, *Astron. Astrophys.* **139**, 477.

SECAM – A SEEING-CONTROLLED CAMERA FOR HIGH-RESOLUTION PHOTOGRAPHS*

WOLFHARD SCHLOSSER

Astronomisches Institut der Ruhr-Universität, Bochum, F.R.G.

(Received 1 October, 1984)

Abstract. The basic design of a seeing-controlled camera (SECAM) is described. The camera is currently under construction, and will be used in combination with the 61 cm telescope of the Ruhr-University Bochum at La Silla, Chile.

The angular resolution of an astronomical telescope is limited by the seeing rather than by its optical performance. Since stellar images have diameters rarely below one second of arc, any telescope with an entrance pupil of, say, 20 cm or larger will suffer from atmospheric turbulences.

Quite a lot of ingenious work has been done to overcome this handicap. The techniques employed range from Michelson's stellar interferometer to speckle-interferometry and active optics, to mention only a few. Common to all these techniques is, however, quite a complicated set-up, since they try to restore the resolution limit of the instrument (about $0\overset{''}{.}1$ for a 1.5 m telescope).

On the other hand, at sites of good seeing there are always moments of excellent image definition well below half a second of arc. If these moments are selected, standard photography can be used to fill – at least partly – the gap between a tenth and one second of arc. Such a camera consists basically of a rapid shutter controlled by a system monitoring seeing *and* image motion.

With financial support of the Ministry of Research of North-Rhine–Westphalia a seeing-controlled camera (SECAM) is currently under construction. It will be used in

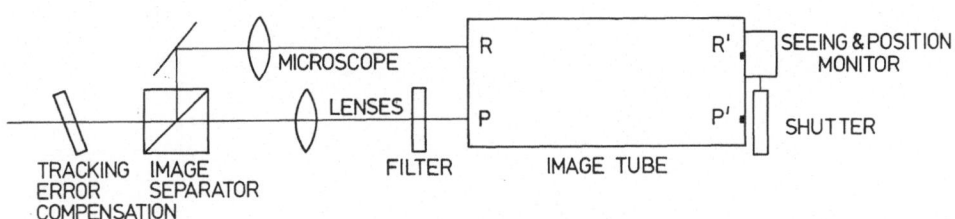

Fig. 1.

* Communication presented at the International Conference on 'Astrometric Binaries', held on 13–15 June, 1984, at the Remeis-Sternwarte Bamberg, Germany, to commemorate the 200th anniversary of the birth of Friedrich Wilhelm Bessel (1784–1846).

Astrophysics and Space Science **110** (1985) 75–76. 0004–640X/85.15

combination with the 61 cm telescope of the Ruhr-University at La Silla, which has a theoretical resolution of $0''2$ at 500 nanometer. Due to its excellent optics, the practical resolution will not be far off.

The basic design of this camera is depicted in Figure 1. Tracking errors of the telescope are compensated by a thick rocking glass-plate moved by two motor-micrometers. The light is then split into two focal planes (image separator). Two microscope lenses focus the program star P and a nearby reference star R onto the photocathode of a 40 mm three-stage electrostatic image tube. Position and central brightness of the image of the reference star R' are continuously monitored. If the accuracy of position and quality of seeing exceed a pre-set threshold, the monitoring system emits an 'open'-command to the shutter, which reacts within milliseconds. The image P' of the program star is then recorded on film.

While SECAM is nearing completion and will be put in operation in early 1985, the optical subsystem has already been tested successfully at the 61 cm Bochum telescope at La Silla. Pictures of R136 and η Carinae were secured with a resolution of half a second of arc or better, which compare favourably with those taken with much larger telescopes.

STATISTICAL MODELS FOR HIPPARCOS BINARIES*

S. SÖDERHJELM

Lund Observatory, Box 43, 5–22100 Lund, Sweden

(Received 13 June, 1984)

Abstract. A statistical model for the binaries in the solar neighbourhood is constructed, using schematic but plausible distribution functions for the semi-major axes, the mass-ratios, and the eccentricities. The model is 'calibrated' to give correct numbers of observed visual binaries and is then used to study the (closer) binaries of relevance for the HIPPARCOS mission. One main purpose is to estimate the influence of astrometric binaries observed by their photocentres. It is found that some 500 such systems (out of the 100 000 stars observed by HIPPARCOS) should show a detectably curved motion, while more than ten times more show an orbital proper motion bias greater than the HIPPARCOS accuracy $0\overset{''}{.}002\,\mathrm{yr}^{-1}$. Also, a non-negligible fraction (1%) of all stars will be binaries with periods close to 1 year causing problems for the parallax determination. The main contribution to these figures is from faint Main-Sequence stars. The period-distribution for the 'resolved' HIPPARCOS binaries is also obtained, and for many of them the periods are fairly short (< 100 yr). Such data as these are to be used as guidelines in the construction of reduction software for the HIPPARCOS observations.

1. Introduction

The HIPPARCOS mission is primarily designed to measure parallaxes and proper motions for *single* stars. As is well known, however, the proportion of binaries and multiple stars is quite high, and it is important to be able to estimate in advance the kind of problems to expect in the reductions from this cause. The present paper aims to derive the statistical distribution of some important binary parameters for typical stellar groups observed by HIPPARCOS.

As is well known, the knowledge of the 'true' distributions of semi-major axes, mass-ratios, and eccentricities (including their variation with mass and age) for the binaries in the solar neighbourhood is very fragmentary. I have chosen, therefore, to make a rather coarse model, using simple, uncorrelated distribution functions. The model output is found to be rather insensitive to the input assumptions, and some useful guidelines for the HIPPARCOS reductions are finally obtained.

2. Model Overview

A classical 'shell'-model is constructed for whole-magnitude steps in apparent magnitude. The number of stars within $m \pm 0.5$ in apparent magnitude and within $M \pm 0.5$ in absolute magnitude is written as

$$N(m, M) = V_{\mathrm{eff}}(m - M)\,\phi(M), \tag{1}$$

---

* Communication presented at the International Conference on 'Astrometric Binaries', held on 13–15 June, 1984, at the Remeis-Sternwarte Bamberg, Germany, to commemorate the 200th anniversary of the birth of Friedrich Wilhelm Bessel (1784–1846).

Astrophysics and Space Science **110** (1985) 77–87. 0004–640X/85.15

where V_{eff} is a volume corrected for the star density decrease in the galactic z-direction and ϕ is a standard luminosity function. The stars in such an $(m - M)$-volume are all assumed to be at the distance $d_{eff}(m - M)$, and V_{eff} and d_{eff} depend on the star distribution model (see below). A proportion F_{bin} (generally $\gtrsim 1$, because triples are counted as 2 binaries, etc.) of the stars are binaries, and these are then assumed to be distributed in semi-major axes (a) and mass-ratio (q) according to the probability density functions $f_a(a)$ and $f_q(a, q)$. Assuming a unique mass-luminosity relation, the q-distribution can be transformed to a Δm-distribution and vice versa. The orbital planes and periastron directions are assumed to be randomly distributed, and we only have to specify finally the probability density for the eccentricities, $f_e(e)$.

In a first 'calibration', we may vary the parameters of the model in order to fit some observed binary statistics. I am using for this purpose the data for visual binaries given by Heintz (1969). After the model has passed this test, it can be used to derive more interesting distributions of parameters observable by HIPPARCOS.

3. Basic Assumptions and Parameters

As for the magnitude system, we will assume that the effective HIPPARCOS-magnitude m_H is a straight mean of Johnson's V and B (unsubscripted m and M refer to m_H and M_H). The luminosity function $\phi_H(M)$ should then be some mean of ϕ_V and ϕ_B, and for simplicity I have used the analytical approximation

$$\phi_H(M) = 0.0030 \times 10^{0.05(M - 1.7)} [1 + 10^{-0.22(M - 1.7)}]^{-2.8}, \tag{2}$$

as can be derived from the expressions for ϕ_V and ϕ_B given by Bahcall and Soneira (1980). The values given by Equation (2) are generally within 50% of the geometric mean of some current ϕ_V and ϕ_B listed by Philip and Upgren (1983).

For the space-distribution, we assume a model galaxy with constant star density in the galactic plane and an exponential decrease with z. We assume also an absorbing layer with $k = 1.0$ mag kpc^{-1} uniformly filling the interval $|z| < 250$ pc. For a given scale-height z_0, we may then derive numerically the effective volumes

$$V_{eff}(z_0, y) = 4\pi \int_0^{\pi/2} d\theta \int_{r(y - 1/2, \theta)}^{r(y + 1/2, \theta)} r^2 \cos \theta \exp(-r \sin \theta / z_0) \, dr, \tag{3}$$

where the distance modulus is related to r by

$$y = 5 \log(r/10) + kr, \qquad r < r_0,$$
$$y = 5 \log(r/10) + kr_0, \qquad r \geq r_0; \tag{4}$$

and where

$$r_0 = 250/\sin \theta. \tag{5}$$

The corresponding effective mean distances are given by

$$d_{eff}(z_0, y) = \frac{\int r \, dV_{eff}}{\int dV_{eff}}. \tag{6}$$

Using the data by Ochsenbein (1983) as a guide, I have used $z_0 = 80$ pc for $-5.5 < M < -0.5$, $z_0 = 130$ pc for $-0.5 < M < 3.5$, and $z_0 = 250$ pc for $M > 3.5$.

For the mass-luminosity relation, I have used the following numerical relations:

$$M_H = 2.11 - 3.81x - 0.98x^2, \qquad 0.568 \le x < 1.4,$$

$$M_H = 5.10 - 14.34x + 8.29x^2, \qquad -0.195 \le x < 0.568, \tag{7}$$

$$M_H = 4.38 - 21.72x - 10.62x^2, \qquad -0.90 \le x < -0.195,$$

$$M_H = 12.98 - 2.60x, \qquad x < -0.90;$$

where x stands for $\log(M/M_\odot)$. This is an almost continuous curve with (almost) continuous first derivative, and for $x > -0.90$ it conforms well to the data for the main sequence given by Allen (1973) or Lacy (1979). (The faint extension is for numerical convenience only and has no influence on the solution.) Because the separate parts are only second order in x, they can be easily inverted to give the masses as function of the absolute magnitudes.

Many different distribution functions for the semi-major axes of binaries have been suggested. In order to keep the number of free parameters small, I have adopted the following form for $f_a(a)$

$$f_a(x) = 0, \qquad x < -2, \quad x > 4.5,$$

$$f_a(x) = 0.22 + 0.061x, \qquad -2 \le x < 0,$$

$$f_a(x) = 0.22, \qquad 0 \le x \le 1.7, \tag{8}$$

$$f_a(x) = 0.35357 - 0.07857x, \qquad 1.7 < x \le 4.5;$$

where x stands for $\log a$ (AU) and where $\int_{-2}^{4.5} f_a(x)\,dx = 1$. In its main form, the distribution (8) approximates the empirical results found by Abt (1979).

The most uncertain part in any discussion of binary statistics is the distribution of mass-ratios. Following the ideas of Abt (1979), I have first assumed different q-distributions for wide and close systems. (The dividing value of a, a_{sep}, is very uncertain, but of the order of 10 AU). For $a < a_{sep}$, I have used the simple power law

$$f_q(q) = 0.35q^{-0.65}, \tag{9}$$

which corresponds to the approximate $N \sim m^{0.35}$ observed by Abt in *logarithmic* mass-bins. For $a > a_{sep}$, Abt finds a distribution of secondary masses which follows the standard ('van Rhijn') luminosity function. The problem with this is that for bright primaries almost all secondaries become invisible faint M-dwarfs, and the binary frequency has to be increased to absurd values. As a partial remedy, one may introduce a 'cut-off' magnitude difference D_m and consider only $\Delta m \,(\equiv m_{sec} - m_{pr}) < D_m$ (~ 8–10). This still overrepresents the large $\Delta m : s$, as can be seen when one compares with the observations as given by Halbwachs (1983) or Herczeg (1984). For $a > a_{sep}$, I, therefore, finally adopted the linear Δm-distribution

$$f_{\Delta m}(\Delta m) = (20 + 11.2\Delta m)/272, \qquad \Delta m = 0 \text{ to } 1,$$

$$f_{\Delta m}(\Delta m) = (10 + 5.6\Delta m)/272, \qquad \Delta m = 1 \text{ to } 8; \tag{10}$$

conforming roughly to the one given by Halbwachs. Such a Δm-distribution gives quite different q-distributions depending on the absolute magnitude of the primary, but it is a useful extreme assumption.

The orbital planes and the periastra are assumed to be randomly oriented, but there remains to specify the distribution of the eccentricities. If we neglect the tidally-influenced period-eccentricity relation, most studies indicate a rather uniform e-distribution. Theoretically, higher eccentricities should be more frequent, and I have used the limiting distributions

$$f_1(e) = 1/0.9 , \qquad e = 0 \text{ to } 0.9 , \tag{11}$$

$$f_2(e) = 2e/0.81 , \qquad e = 0 \text{ to } 0.9 . \tag{12}$$

4. Calibration of the Model

The total star numbers (primaries) for different apparent magnitudes are quite reasonable. (We find, e.g., 7040 stars with $m_H = 6.0$–7.0 and 50 300 with $m_H = 8.0$–9.0.) The first interesting statistical distribution is that for apparent separations and magnitude differences for classical visual binaries. Here we may compare with the statistics for ADS-binaries given by Heintz (1969). The present model gives the two-dimensional distribution over α (angular semi-major axis) and Δm, while Heintz gives only the marginal distribution over α. Heintz estimates, however, a completeness limit at about

$$0.22\Delta m - \log \alpha('') = 0.75 \tag{13}$$

and our present data may thus be transformed for comparison with his.

The only free parameters in the model are the total binary frequency F_{bin} and the dividing value a_{sep} for the q-distributions. One reasonable model (A) has $a_{sep} = 10$ AU and $F_{bin} = 1.0$. Small shifts of a_{sep} have very little influence, but in order to have an extreme case, the 'close' q-distribution was used for *all* systems ($a_{sep} = \infty$). Surprisingly, the Heintz statistics could be equally well satisfied with this model (B), at $F_{bin} = 1.8$. (This high F_{bin} is due to the large number of small-q systems given by Equation (9).) Table I shows the calculated numbers of binaries and Heintz's observed numbers for three different intervals of apparent magnitude.

In view of the uncalibrated magnitude-scale, the agreement is as good as can be expected. The only obvious systematic trend is the low number of close systems predicted by model A.

5. Detection of Binaries by HIPPARCOS

The wide variety of possible binary configurations (separations, magnitude differences, orbital motion) makes it difficult to generalize, but as a starting point we may use the data given by Lindegren (1982). For relatively *fixed* secondaries, he derives detection limits in the (sep, Δm)-plane that may be realistic for the brighter pairs. (We must have a cut-off at $m_{sec} \sim 13$–14 which certainly precludes observation of systems with, e.g.,

TABLE I

Number of binaries for different separations and magnitudes

$\log \alpha\,('')$	$m = 6–7$			$m = 7–8$			$m = 8–9$		
	Obs.	A	B	Obs.	A	B	Obs.	A	B
− 0.5	51	18	33	109	48	83	338	125	202
− 0.3	65	37	70	119	98	170	443	253	408
− 0.1	63	53	92	139	143	230	485	361	540
+ 0.1	89	71	103	171	189	264	529	470	620
+ 0.3	108	86	110	228	232	284	647	577	660
+ 0.5	96	102	110	233	273	283	685	669	668
+ 0.7	117	118	109	241	306	271	721	746	651
+ 0.9	113	126	102	207	330	262	634	794	610
+ 1.1	122	133	96	211	342	233	517	728	463

Model A: $a_{\text{sep}} = 10$ AU, $F_{\text{bin}} = 1.0$, Model B: $a_{\text{sep}} = \infty$, $F_{\text{bin}} = 1.8$.

$\Delta m = 4–5$ for a 12th magnitude primary.) A conservative straight-line approximation similar to the Heintz criterion (13) gives the following conditions for *possibly* resolved pairs

$$0.22\Delta m - \log \alpha < 1.3 , \tag{14}$$

$$\Delta m < 13.5 - m_{\text{pr}} .$$

Systems outside these limits are observed as single (moving) stars. (For separations greater than some $0.''5$, the primary is observed, otherwise the photocentre.) One primary goal for the present investigation is to estimate how many of these astrometric pairs that may be detected from their curved motions, and also how the undetected ones bias the proper motions.

For orbital periods above some 5 years, the observations of an astrometric binary will be distributed along a curved arc, as a measure of curvature I have used the vector

$$\mathbf{s} \equiv \mathbf{r}[(t_1 + t_2)/2] - [\mathbf{r}(t_1) + \mathbf{r}(t_2)]/2 , \tag{15}$$

where $\mathbf{r}$ is the radius vector to the photocentre in units of its semi-major axis α_{ph}, and the mission extends from t_1 to t_2. As a fair approximaton, we may then assume that perpendicular to some direction, the deviations of the path from a straight line can be written

$$\Delta x = -\tfrac{2}{3}s + 4s(t - t^2) \tag{16}$$

where t_1 is now equal to 0 and $t_2 = 1$. For uniformly spaced observations, the variance of Δx is then $4s^2/45$, and assuming the HIPPARCOS scans to be randomly oriented relative to the 'reference' direction, we find an extra variance in the positions along the scan circle (the 'abscissae') equal to one half of this, viz.,

$$v_{\text{absc}}^{\text{rel}} = 2s^2/45 . \tag{17}$$

Different binaries of course give widely different s^2-values, and it is necessary to investigate the expected distribution of s^2 as function of orbital period and eccentricity. This was done numerically for a large number of binaries (40 000) of given period and eccentricity, but with uniform distributions over orientation angles and mean anomalies. The combined distributions for the two different e-distributions differ relatively little, and in each case we find a rather flat distribution with 'wings' extending about 1.5 dex in s^2. With a spacing in $\log P$ of 0.1 dex, it is easy to interpolate a numerical s^2-distribution at any $P > 5$ yr.

For very short orbital periods, the entire orbit is covered more or less uniformly. For this ideal uniform case, and with scans in random directions, we now have something like

$$v^{\text{rel}}_{\text{absc}} = \tfrac{1}{2} |\mathbf{r} - \langle \mathbf{r} \rangle|^2 . \tag{18}$$

The distribution of this quantity was again calculated numerically, and the only lacking part of the distribution of $v^{\text{rel}}_{\text{absc}}$ is the difficult region with $P \sim 1$–5 yr. Somewhat arbitrarily, I have chosen to use the 'short-P' distribution for all $P < 2.5$ yr, and to interpolate linearly between the 2.5 yr and 5 yr-distributions for a P in this range.

For the proper motions we proceed similarly. For $P > 5$ yr, we let

$$\mathbf{p} \equiv (\mathbf{r}(t_2) - \mathbf{r}(t_1))/(t_2 - t_1) \tag{19}$$

and derive the distribution of p^2 as that of s^2 before. The short-P limit is now assumed to be zero, and the 2.5–5 yr data are interpolated as before. (It is important to note that s^2, $v^{\text{rel}}_{\text{absc}}$, and p^2 are so far given in units of the angular semi-major axis of the photocentre, α_{ph}.)

6. The HIPPARCOS Astrometric Binaries

The general binary star model predicts the number of systems in bins of e.g., $\log \alpha$, $\log P$, q, and Δm. The semi-major axis of the photocentre is given by

$$(\alpha_{ph}/\alpha) = q/(1 + q) - (1 + 10^{0.4\Delta m})^{-1} , \tag{20}$$

and by this equation we may transform the original multi-variate distribution to one over the primary variables $\log \alpha_{ph}$ and $\log P$. For each bin in this distribution, we have a distribution of $v^{\text{rel}}_{\text{absc}}$ and p^2, and by a suitable convolution, we finally obtain univariate distributions of v_{absc} and v_{pm}.

The extra abscissa variance may be detected when it exceeds the observational scatter by a certain factor. For a standard 1% χ^2-criterion, the detection limit is at about $\log v_{\text{absc}} = -5.0$ (arcsec^2) for the brighter stars, increasing to maybe $\log v_{\text{absc}} = -4.5$ for an 11th mag. primary. Table II gives the expected numbers of stars with $\log v_{\text{absc}} > -5.05$ for different apparent magnitudes and for the A and B models defined above. Postscripts 1 and 2 refer to the eccentricity-distributions (11) and (12), which are seen to have little influence on these results. Altogether, relatively few stars are

TABLE II

Numbers of astrometric binaries with detectably curved motion of their photocentres

m Model	2–3	3–4	4–5	5–6	6–7	7–8	8–9	9–10	10–11	2–9
$A1$	3	7	17	34	61	96	138	176	204	357
$A2$	2	7	16	33	57	90	128	161	186	333
$B1$	5	13	31	62	109	172	248	316	367	642
$B2$	5	13	29	59	103	161	229	290	334	601

detectable by their curved motion, and by separating the absolute magnitudes, it may be shown that they are mostly F–K dwarfs.

As for the proper motion variance (v_{pm}) due to binaries, it affects sensibly a larger proportion of stars. Table III shows the percentages of stars with $\sqrt{v_{pm}} > 0\rlap{.}''002 \text{ yr}^{-1}$ at different apparent magnitudes. As a comparison, Lindegren (1978) obtained the round figure 10% for $m_{pg} = 9$ stars (from a more limited study).

TABLE III

Relative numbers of stars (%) with $\sqrt{v_{pm}} > 0\rlap{.}''002 \text{ yr}^{-1}$

m Model	2–3	3–4	4–5	5–6	6–7	7–8	8–9	9–10	10–11	2–9
$A1$	9.9	10.7	11.0	10.7	10.1	8.7	7.0	5.4	3.9	7.9
$A2$	9.5	10.2	10.4	10.1	9.4	8.1	6.5	4.9	3.6	7.3
$B1$	16.1	16.0	15.2	13.8	12.1	10.1	7.9	5.9	4.3	9.1
$B2$	15.4	15.2	14.4	13.0	11.3	9.3	7.3	5.5	3.9	8.4

It is instructive to separate the stars according to absolute magnitude. Table IV gives the resulting percentages for apparent magnitude $m = 8$–9. For the faintest stars, the proportion of large 'binary' proper motions is high, but of course their 'true' proper motions are also higher.

TABLE IV

Relative number of stars (%) showing various degrees of proper motion variance. Apparent magnitude interval $m = 8$–9.

$\sqrt{v_{pm}}$ \ M	-1.5 to $+0.5$		$+0.5$ to $+2.5$		2.5 to 5.5		>5.5	
	A_1	B_1	A_1	B_1	A_1	B_1	A_1	B_1
$>0\rlap{.}''002 \text{ yr}^{-1}$	0.8	0.9	6.6	7.1	14.3	16.3	16.5	24.5
$>0\rlap{.}''005 \text{ yr}^{-1}$	–	–	0.3	0.3	5.0	5.7	12.5	15.7
$>0\rlap{.}''010 \text{ yr}^{-1}$	–	–	–	–	0.8	1.0	6.8	8.4
$>0\rlap{.}''020 \text{ yr}^{-1}$	–	–	–	–	0.0	0.0	1.8	2.5
$>0\rlap{.}''050 \text{ yr}^{-1}$	–	–	–	–	–	–	0.0	0.1

From the distribution of astrometric binaries in the $(\log \alpha_{ph}, \log P)$-plane we may also estimate the number of cases with P close to one year and $\alpha_{ph} > 0\overset{\prime\prime}{.}002$ where the parallaxes may become faulty. Table V gives these numbers for the $\log P$ interval -0.1 to $+0.1$. A division by luminosity shows again that the percentages are higher for fainter absolute magnitudes.

TABLE V

Numbers of stars where the parallax error may exceed 50%

Model \ m	2–3	3–4	4–5	5–6	6–7	7–8	8–9	9–10	10–11	2–9
A	–	3	10	24	56	116	218	342	562	427
%		(1)	(1.2)	(1.0)	(0.8)	(0.6)	(0.4)	(0.3)	(0.2)	(0.5)
B	2	6	17	43	101	209	391	661	1013	769
%	(3)	(2.4)	(2.0)	(1.7)	(1.4)	(1.1)	(0.8)	(0.5)	(0.4)	(1.0)

All the figures in Tables II–V are for an assumed 'Main-Sequence' luminosity function. As shown by Halbwachs (1985), we should expect also a rather high proportion of degenerate components to apparently single Main-Sequence stars, and I have made some preliminary calculations to include this in the model. Using the same a- and e-distributions as for the Main-Sequence binaries, and assuming that a typical degenerate component has a mass equal to $0.8\,M_{\odot}$, I find that the figures in Tables II–V should be increased by 20–50% (higher values for the A model) for a 10% fraction of degenerates. The uncertainty due to this cause is thus of the same typical magnitude as the difference between the A and B models.

7. The Resolved HIPPARCOS Binaries

In Section 6 we obtained the statistics for binaries observed as single stars by HIPPARCOS. The resolution criterion (14) is necessary, but not sufficient, because Lindegren's (1982) study implicitly assumed components *fixed* relative to each other. The present model allows us to study the period-distribution among the 'tentatively resolved' pairs. There are many combinations of separations, magnitude differences, and absolute and apparent magnitudes to be considered, but in Table VI are given some results for the 'typical' $m = 8$–9 interval. The numbers are for all 'new' (undetected by the Heintz criterion, but not necessarily by the time of the HIPPARCOS mission) binaries, regardless of Δm. As expected, the close pairs also show the more rapid motion, and there will obviously be detection problems for some systems in the upper left corner of Table VI. (For apparently brighter pairs, the periods are shifted to even shorter values, aggravating the problems.) A separation according to absolute magnitude shows that again the faint nearby pairs are the problematic ones. Table VII gives the star numbers in column 5 of Table VI ($\log \text{sep} = -0.5$), and similar distributions are obtained for other separations.

TABLE VI

'Resolved' binaries according to separation and period for apparent magnitude $m = 8$–9

log sep (")	-1.1		-0.9		-0.7		-0.5		-0.3		-0.1		+0.1		+0.3	
log P (yr)	A	B	A	B	A	B	A	B	A	B	A	B	A	B	A	B
<1	13	20	8	6	0	2	–	–	–	–	–	–	–	–	–	–
1.0–1.5	61	85	44	51	22	29	6	6	–	–	–	–	–	–	–	–
1.5–2.0	137	153	170	210	130	139	62	51	25	16	4	7	0	2	–	–
2.0–2.5	43	82	166	261	270	343	192	211	116	89	38	26	7	5	0	1
2.5–3.0	–	–	16	40	114	204	254	300	264	216	122	72	35	30	3	3
3.0–3.5	–	–	–	–	5	7	51	70	190	179	184	157	92	61	7	0
3.5–4.0	–	–	–	–	–	–	–	–	12	16	67	59	77	68	19	35
>4	–	–	–	–	–	–	–	–	–	–	4	7	9	9	3	15

TABLE VII

Numbers of binaries according to absolute magnitude and period for $m = 8-9$, separation $= 0\overset{''}{.}25-0\overset{''}{.}40$

$\log P$ \\ M	-5.5 to -1.5		-1.5 to $+0.5$		0.5 to 2.5		2.5 to 5.5		>5.5	
	A	B	A	B	A	B	A	B	A	B
1.0–1.5	–	–	–	–	–	–	–	–	6	8
1.5–2.0	–	–	–	–	–	–	58	50	3	2
2.0–2.5	–	–	–	–	86	106	104	102	–	–
2.5–3.0	0	3	107	140	159	159	–	–	–	–
3.0–3.5	23	43	29	25	–	–	–	–	–	–
3.5–4.0	–	–	–	–	–	–	–	–	–	–

8. Conclusions

The statistical model described above is thought to contain the general features of the 'real' distribution of binaries. From the numerical results we may tentatively conclude:

(1) Variation of the assumed input distributions give relatively small (less than a factor two) changes in the output. Conversely, the HIPPARCOS observations will not *greatly* increase our knowledge of binary statistics.

(2) The Abt model, with different distributions of the mass-ratio (q) for wide and close binaries, will be difficult to distinguish from models with a single distribution.

(3) Only a few hundred astrometric binaries are detectable by HIPPARCOS on account of their curved paths on the sky. Many more (of the order of 10%) will show orbital proper motions of their photocentres greater than $0\overset{''}{.}002$ yr^{-1}.

(4) About 0.5–1% of the stars are binaries with photocentric semi-major axes greater than $0\overset{''}{.}002$ *and* periods in the range 0.8–1.25 yr where the parallax determination is influenced.

(5) A large number of binaries *may* be resolved (observed as two separate components) by HIPPARCOS. Many of them show sensible orbital motion, however, and the reduction methods have to take this into account.

Acknowledgement

This work is supported by the Swedish Board for Space Activities.

References

Abt, H. A.: 1979, *Astron. J.* **84**, 1591.
Allen, C. W.: 1973, *Astrophysical Quantities*, 3rd edition, Athlone Press, London.
Bahcall, J. N., Soneira, R. M.: 1980, *Astrophys. J. Suppl.* **44**, 73.
Halbwachs, J. L.: 1983, *Astron. Astrophys.* **128**, 399.
Halbwachs, J. L.: 1985, *Astrophys. Space Sci.* **110**, 159 (this issue).
Heintz, W. D.: 1969, *J. Roy. Astron. Soc. Can.* **63**, 275.
Herczeg, T.: 1984, *Astrophys. Space Sci.* **99**, 29.

Lacy, C. H.: 1979, *Astrophys. J.* **228**, 817.
Lindegren, L.: 1978, *Colloq. on European Space Astrometry, Padova*, p. 117.
Lindegren, L.: 1982, *Colloq. on the Scientific Aspects of the Hipparcos Space Astrometry Mission, Strasbourg*, ESA SP-177, p. 147.
Ochsenbein, F.: 1983, *Astron. Astrophys.* **118**, 197.
Philip, A. G. D. and Upgren, A. R.: 1983, *IAU Colloq.* **76**, L. Davis Press, New York, p. 471.

DE MEDIA AEQUINOCTIORUM PRAECESSIONE ATQUE DE NUTATIONE STELLAE SS433*

REMO RUFFINI and SONG DOO JONG

Department of Physics, University of Rome 'La Sapienza', Italy

(Received 15 October, 1984)

Abstract. On the basis of Bessel's classical papers *De Media Aequinoctiorum Praecessione* and *De Nutatione* compact formulae are given to interpret the observed variations in the shifted lines of SS433 in terms of nutational and precessional effects of a disk in a binary system.

At the very basis of the dynamics of the Earth motion within the solar system are motion components giving rise to precession and nutation. These effects originate in the quadrupole moment of the Earth, in the spin of the Earth and in the orbital angular momentum of the Earth.

F. W. Bessel has been the first to give in the articles *De Media Aequinoctiorum Praecessione* and *De Nutatione* and analytic treatment of these effects and the often quoted 'Besselian year' together with the Besselian star constants and Besselian star numbers give the method of observing the nutational effects on the background of the 'fixed stars' in the celestial spheres.

Paradoxically in a different context one of the most important relativistic systems in the Galaxy recently discovered, SS433, the basic dynamics is still dominated by precessional and nutational effects.

After recollecting some of the main features of Bessel's idea we like to draw attention to the observational behavior of the SS433. The idea of Bessel may be still used in the understanding of this most unique system in our Galaxy which promises to give a basic new understanding in relativistic astrophysics.

It is well known that the motion of the Earth in the solar system relative to its center of mass is determined by the precession and the nutation. In terms of the astronomical celestial coordinate systems, the equator system and the ecliptic system, we can describe the variations of coordinates at any instant due to precession and nutation. If we define the inclination of the equator system to the fixed ecliptic system at an epoch as ε, obliquity at that date, and the displacement of descending node on the fixed ecliptic as Ψ from the fixed mean equinox of a given epoch, then the variations of these angles due to the displacement of the equatorial circle with respect to the ecliptic circle (Laplace,

* Communication presented at the International Conference on 'Astrometric Binaries', held on 13–15 June, 1984, at the Remeis-Sternwarte Bamberg, Germany, to commemorate the 200th anniversary of the birth of Friedrich Wilhelm Bessel (1784–1846).

1825; Woolard and Clemence, 1966)

$$\varepsilon = \varepsilon_0 + S_\varepsilon(t) + \Delta\varepsilon ,$$

$$\Psi = S_\Psi(t) + \Delta\Psi ; \tag{1}$$

where ε_0 is the mean obliquity at the epoch, $S_\varepsilon(t)$ and $S_\Psi(t)$ are a secular function of time due to precession in longitude and precession in obliquity, respectively, given by Laplace (1825)

$$S_\Psi(t) = 50''{.}287\,60t - 0''{.}000\,121\,974\,5t^2 ,$$

$$S_\varepsilon(t) = 0.000\,009\,842\,33t^2 \tag{2}$$

and $\Delta\Psi$ and $\Delta\varepsilon$ are periodic terms due to nutation (Bessel, 1875) given by

$$\Delta\Psi = -16''{.}783\,32 \sin\Omega + 0''{.}202\,09 \sin 2\Omega -$$

$$- 1''{.}335\,89 \sin 2\Omega_\odot - 0''{.}201\,28 \sin 2\Omega_{\mathbb{C}} ,$$

$$\Delta\varepsilon = +8''{.}977\,07 \cos\Omega - 0''{.}087\,73 \cos 2\Omega + \tag{3}$$

$$+ 0''{.}579\,90 \cos 2\Omega_\odot + 0''{.}087\,38 \cos 2\Omega_{\mathbb{C}} ;$$

in which Ω denotes the mean longitude of the lunar node and the $\Omega_{\mathbb{C}}$ and $\Omega_\odot$ are the mean longitude of the Moon and the Sun.

Due to these precessional and nutational variations of the equator system relative to the fixed ecliptic system at a given epoch, the position of celestial body in the equatorial system, is given by a set of angles α (right ascension) and δ (declination) changing with time. Bessel firstly did calculate these variations of position of celestial bodies (Bessel, 1875). Practically a position (α, δ) referred to the real equinox and equator at any particular data is usually obtained by first determining the mean position (α_0, δ_0) at the beginning of the Besselian year in which the date lies, then adding the further precession and nutation at the date of derivation. According to Bessel, the variations due to nutation and precessions, up to first order, are (Bessel, 1875; Woolard and Clemence, 1966)

$$\Delta\alpha = n\left\{\tau + \frac{\Delta\Psi}{\Psi'}\right\}\left(\frac{m}{n} + \sin\alpha_0 \tan\delta_0\right) + \lambda'\frac{\Delta\Psi}{\Psi'} + \cos\alpha_0 \tan\delta_0 \, \Delta\varepsilon ,$$

$$\tag{4}$$

$$\Delta\delta = n\left\{\tau + \frac{\Delta\Psi}{\Psi'}\right\} \cos\alpha_0 + \sin\alpha_0 \, \Delta\varepsilon ;$$

where Ψ' is the annual rate of luni-solar precession on the fixed ecliptic at the beginning of the Bessel year, λ' is the annual rate of planetary precession in right ascension, $m = \Psi' \cos\varepsilon_0 - \lambda'$, $n = \Psi' \sin\varepsilon_0$ for $\tau = 0$, and τ is the fraction of the year intervening between the Bessel year and time of duration. Referring to a, b and a', b' as

$$a = \frac{m}{n} + \sin\alpha_0 \tan\delta_0 , \qquad b = \cos\alpha_0 \tan\delta_0 , \qquad a' = +\cos\alpha_0 ,$$

$$b' = -\sin\alpha_0 , \tag{5}$$

we know that these factors are so nearly constant for any particular star that the same values can be used for each star over an extended interval of time and, therefore, we call them – in honour of Bessel – the Besselian star constants. We can define other interesting factors known as Besselian star numbers, which are mainly dependent on the precessional and nutational motions of the reference circles, as

$$A = n\left(\tau + \frac{\Delta\Psi}{\Psi'}\right), \qquad B = -\Delta\varepsilon, \qquad E = \lambda'\frac{\Delta\Psi}{\Psi'} \; ; \qquad (6)$$

where the terms in τ give the effect of precession and others due to the nutation in longitude and obliquity. So much for Bessel's work on the planetary system.

In the extra-solar system we have observational evidences of solar type precessional and nutational variations. One – the most interesting and important object – is SS433. The SS433 is a stellar system which has stimulated a great many observational and theoretical studies in recent years because of the numerical behavior of its Doppler-shifted emission line features. From the careful interpretation of observed spectra of SS433, have been obtained the following properties of SS433 (see, e.g., Fang et al., 1981).

(1) The Doppler shifts of emission lines are corresponding to the velocities $v \simeq 0.26c$.

(2) The modulation of the Doppler shifts is cyclic with periods of ~ 164 days for long-term (Margon et al., 1979) and $\sim 6^{d}\!28$ and $\sim 5^{d}\!84$ for short-term (Newsom and Collins, 1980; Katz et al., 1982; Mammano et al., 1983; Cogotti et al., 1985) variations.

These observational conclusions lead us to consider that the SS433 is a unique object in our Galaxy which has a relativistic effect with periodic changes.

In addition to the above properties, the spectroscopic data (Crampton et al., 1980) and photometric data (Margon et al., 1980) have shown that the SS433 is a member of a close binary system with a period of about $13^{d}\!08$.

If we know these properties of SS433, there are variations of models to explain theoretically the extraordinary phenomena of SS433. Among these, two basic geometries have been advanced (Milgrom, 1979): in one (the twin jet model), the emission occurs from two highly collimated beams moving at a speed $v \simeq \frac{1}{4}c$ away from a central object presumably a gravitationally collapsed object. Assuming the jets are normal to an accretion disk which has assumed precession in the binary system, we can explain the observed precessional and nutational effect (Katz et al., 1982); in the other (the ring model), the emission occurs from two opposite regions of a ring in the accretion disk orbiting a fully graviationally collapsed object at a distance of $r_0 \simeq 45M_1$. In this last model the shifts originate from general relativistic effects: a combination of gravitational and Doppler shifts (Fang Li Zhi and Ruffini, 1979; Ruffini and Stella, 1980). In the following paragraphs we would like to discuss the precessions and nutations of SS433 in the basis of ring model and the jet model.

In the case of SS433 the quadrupole terms giving rise to the coupling with the angular momentum of system (corresponding to the quadrupole momentum of the Earth in solar system) is due to the accretion disk surrounding the gravitationally collapsed object.

Such disk of size $R \sim 10\,R_\odot$ is assumed to exist both in the jet model and in the ring model. The differences in each case, of course, are the origin of the shifted emitting region: in the plane of the disk in the ring model and in the jet beam normal to the disk in the jet model. This different emitting point correspondingly gives rise to different parameters of the binary system (see below).

As in the case of the solar system we define for binary system angles ϑ as the inclination of accretion disk with respect to the orbital plane (corresponding to the angle of obliquity) and Φ for the displacement of node along the circle of the orbital plane. Then the variations of ϑ and Φ due to precessions and nutations analogous to the solar system (see Equation (1)) (Cogotti $et\ al.$, 1984; Ruffini and Song, 1984) are

$$\vartheta \simeq \vartheta_0 + \frac{\Omega_E}{2(\omega - \Omega_p)}\ \tan\vartheta_0 \cos[2(\omega - \Omega_p)t - \Lambda_1]\,,$$

$$(7)$$

$$\Phi \simeq \Omega_p t + \frac{\Omega_E}{2(\omega - \Omega_p)}\ \sin[2(\omega - \Omega_p)t - \Lambda_2]\,;$$

where Ω_p is the assumed total precession of the ring, Λ_1 and Λ_2 are phases, ϑ_0 is the mean inclination angle at a fixed initial line, and Ω_E is the precessional angular velocity due to the quadrupole moment of a ring in the tidal field of companion star given by

$$\Omega_E \simeq -\frac{3}{4}\left(\frac{\omega}{\omega_0}\right)\omega\ \frac{M_2}{M_1 + M_2}\ \cos\vartheta_0\,,$$

$$(8)$$

in which ω_0 is angular velocity of a ring and M_1 and M_2 are the mass of compact object and companion star in the binary system.

As Bessel calculated the variation of position angles of equator system due to precession and nutation of the Earth, these variation of angle parameters of the binary system will create the irregularities to the observed emission lines spectrum. Therefore, substituting the angles given in Equation (7) into the formula of Doppler shift (Ruffini and Stella, 1980) and choosing up to the first order, we have the red (blue) shifts of emission lines (Cogotti $et\ al.$, 1984).

$$1 + Z_\pm \simeq 1 + Z_0^\pm \pm A_1 \cos(\Omega_p t + \varphi_1) \mp A_2 \cos(2(\omega - \Omega_p)t + \varphi_2) \pm$$
$$\pm A_3 \cos((2\omega - \Omega_p)t + \varphi_3) \mp A_4 \cos((2\omega - 3\Omega_p)t + \varphi_4)\,,$$

$$(9)$$

where $1 + Z_0^\pm$, A_i, $i = 1, \ldots, 4$, are constants given in Table I and φ_i, $i = 1, \ldots, 4$ are phases.

A comparison of Equation (9) with the experimental data (Frasca $et\ al.$, 1983) allows us to determine the parameters of the binary system such as the masses of the binary system (Cogotti $et\ al.$, 1985) $M_1, M_2 \lesssim 50\,M_\odot$, the inclination angles ϑ_0 as well as other characteristic parameters of the emitting region (see Table II). Future works in this field may very well lead to the confirmation of one of the two possible models discussed above.

TABLE I

We reported here the theoretical amplitudes of Doppler shifts for the ring model and jet model.
$\Gamma = (1 - 3M_1/r_0)^{1/2}$, $\gamma = 1/(\sqrt{1 - \beta^2})$, and $\Delta\Lambda = \Lambda_1 - \Lambda_2$.

Model Amp.	Ring	Jet
$1 + Z_0^{\pm}$	$\Gamma[1 \pm (M_1/r_0)^{1/2} \cos i \sin\vartheta_0]$	$\gamma(1 \pm \beta \cos i \cos\vartheta_0)$
A_1	$\Gamma(M_1/r_0)^{1/2} \sin i \cos\vartheta_0$	$\gamma\beta \sin i \sin\vartheta_0$
A_2	$\Gamma(M_1/r_0)^{1/2} \dfrac{\Omega_E}{2(\omega - \Omega_p)} \cos i \sin\vartheta_0$	$\gamma\beta \dfrac{\Omega_E}{2(\omega - \Omega_p)} \cos i \cos\vartheta_0 \tan^2\vartheta_0$
A_3	$\frac{1}{2}\Gamma(M_1/r_0)^{1/2} \dfrac{\Omega_E}{2(\omega - \Omega_p)} \sin i \cos\vartheta_0 \times$ $\times [1 + \tan^4\vartheta_0 - 2\cos\Delta\Lambda \tan^2\vartheta_0]^{1/2}$	$\dfrac{1}{\sqrt{2}} \gamma\beta \dfrac{\Omega_E}{2(\omega - \Omega_p)} \sin i \sin\vartheta_0(1 + \cos\Delta\Lambda)^{1/2}$
A_4	$\frac{1}{2}\Gamma(M_1/r_0)^{1/2} \dfrac{\Omega_E}{2(\omega - \Omega_p)} \sin i \cos\vartheta_0 \times$ $\times [1 + \tan^4\vartheta_0 + 2\cos\Delta\Lambda \tan^2\vartheta_0]^{1/2}$	$\dfrac{1}{\sqrt{2}} \gamma\beta \dfrac{\Omega_E}{2(\omega - \Omega_p)} \sin i \sin\vartheta_0(1 + \cos\Delta\Lambda)^{1/2}$

TABLE II

The values of the parameters of SS433 system are reported and compared. For this table we have used $\omega \simeq 2\pi/13\overset{d}{.}08$ and $\Omega_p \simeq 2\pi/163\overset{d}{.}34$.

Model Parameter	Ring	Jet
(i, ϑ_0)	$(65\overset{\circ}{.}7, 54\overset{\circ}{.}0)$	$(78\overset{\circ}{.}3, 18\overset{\circ}{.}4)$
r_0	$45\,M_1$	–
v/c	–	0.26
$\Delta\Lambda$	$127\overset{\circ}{.}3$	$60°$
Ω_E	$2\pi/83\overset{d}{.}8$	$2\pi/55\overset{d}{.}3$

Bessel's pioneerwork on the understanding of the coupling of the quadrupole moment of the Earth to the total angular momentum of the planetary motion and of lunisolar system provided a detailed rules for measuring this effect out of the apparent motion of the fixed stars. It is gratifying to that this same treatment of Bessel's can still be used in its essential form for the understanding of this most unique relativistic system in our Galaxy – SS433 – seen, this time, from the faraway planet Earth.

References

Bessel, F. W.: 1875, in R. Engelmann (ed.), *Bessel Abhandlungen*, Vol. I, Leipzig.

Cogotti, R., Ruffini, R., and Song, D. J.: 1985, *Astron. Astrophys.*, (in press).

Crampton, D., Cowley, A. R., and Hutchings, J. B.: 1980, *Astrophys. J.* **235**, L31.

Fang Li Zhi and Ruffini, R.: 1979, *Phys. Letters* **B86**, 193.

Fang Li Zhi, Ruffini, R., and Stella, L.: 1981, *Vistas Astron.* **25**, 185.

Frasca, S., Ciatti, F., and Mammano, A.: 1983, Preprint.

Katz, J. I., Anderson, S. E., Margon, B., and Grandi, S. A.: 1982, *Astrophys. J.* **260**, 780.

Laplace, P. S.: 1825, *Traité de la Mécanique Céleste*, Vol. III, p. 158.

Mammano, A., Margoni, R., Ciatti, F., and Cristiani, S.: 1983, *Astron. Astrophys.* **119**, 153.

Margon, B., Ford, H. C., Grandi, S. A., and Stone, R. P. S.: 1979, *Astrophys. J.* **233**, L63.

Margon, B., Grandi, S. A., and Downes, R. A.: 1980, *Astrophys. J.* **241**, 306 (see also Anderson, S. F., Margon, B., and Grandi, S. A.: 1983, *Astrophys. J.* **269**, 605).

Meritt, D., and Petterson, J. A.: 1980, *Astrophys. J.* **236**, 255.

Milgrom, M.: 1979, *Astron. Astrophys.* **76**, L3.

Newsom, G. H. and Collins, G. W.: 1980, *IAU Circ.*, No. 3459.

Ruffini, R. and Stella, L.: 1980, *Nuovo Cimento Letters* **27**, 529.

Ruffini, R., and Song, D. J.: 1984, *Astrophys. Space Sci.* **99**, 319.

Woolard, E. W. and Clemence, G. M.: 1966, *Spherical Astronomy*, Academic Press, New York.

RELATIVISTIC ASTROMETRY*

MICHAEL SOFFEL, JOACHIM SCHASTOK, and HANNS RUDER

Lehrstuhl für theor. Astrophysik, Universität Tübingen, F.R.G.

and

MANFRED SCHNEIDER

Sonderforschungsbereich, Technische Universität München, F.R.G.

(Received 25 July, 1984)

Abstract. Concepts of relativistic astrometry – such as Weyl's stellar compass or the concept of 'flat-space plus forces' – are discussed. To visualize effects from light deflection pictures showing the stellar sky as seen from the vicinity of a strongly gravitating source are presented.

The rapid advance in VLBI and space-borne astrometric techniques with the capability to determine relative positions for objects with small angular separations with accuracies of a few milliarcseconds will change the character of astrometry significantly in the near future: since at this level of accuracy effects from light deflection in the gravitational field of the Sun, Jupiter and Saturn have to be considered for the interpretation of astrometric measurements and the determination of quasi-inertial frames astrometry will become a theory over curved space-time.

As is well known, the angle of light deflection at the limb of the Sun according to Einstein's theory of gravity amounts to $1''.75$ and decreases with increasing photon impact parameter like $1/r$. What is less appreciated is the fact that for impact parameters ~ 1 AU this effect still yields

$$2\Delta\varphi \simeq 1''.75 \, R_\odot/\text{AU} \simeq 1''.75 \times 5 \times 10^{-3} \simeq 8 \times 10^{-3}{''}.$$

Thus, for light rays incident at about $90°$ from the Sun the angle of light deflection still amounts to $4 \times 10^{-3}{''}$ and should be detectable by HIPPARCOS or the space telescope.

Let us consider the solar system as being isolated – i.e., space time approaches flat Minkowski space 'far away' from the solar system but still in a regime where effects from Hubble expansion are not appreciable. Figure 1a shows a representation of a 3-dimensional isolated space-time where a conformal transformation of the metric brought the various asymptotic ('far away') parts ($I^{\pm,0}$, $\mathscr{I}^{\pm}$) of space-time into a finite location. In such a (Penrose) diagram light rays must originate from $\mathscr{I}^-$ (past null infinity) in the distant past and will end in $\mathscr{I}^+$ future null infinity) in the distant future. Pedestrians moving with velocity less than the speed of light originate from I^- and will end in I^+. Spacelike geodesics both originate and end in I^0. Let P be any point of this space-time $\mathscr{M}$ (see Figure 1b) and κ null-vector in P. Then there will be exactly one null-geodesic

---

* Communication presented at the International Conference on 'Astrometric Binaries', held on 13–15 June, 1984, at the Remeis-Sternwarte Bamberg, Germany, to commemorate the 200th anniversary of the birth of Friedrich Wilhelm Bessel (1784–1846).

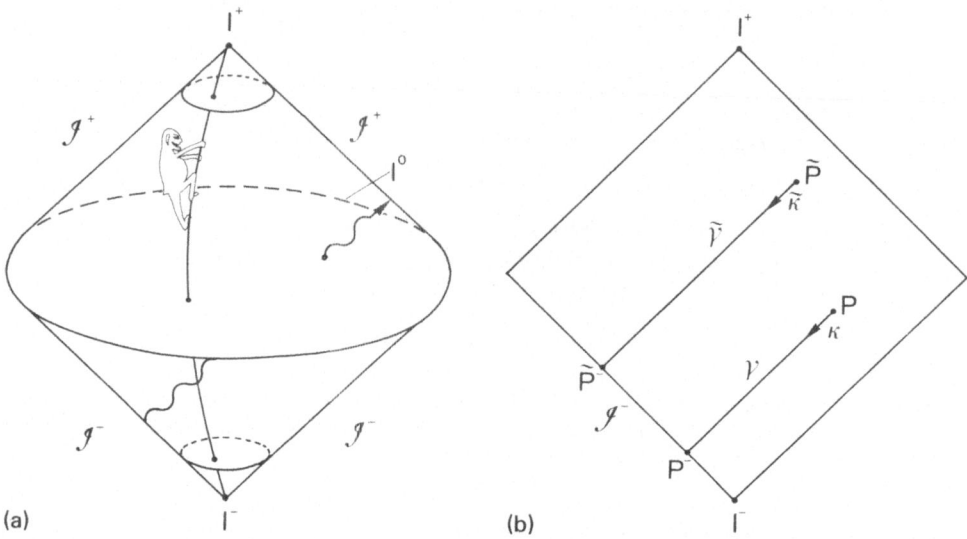

Fig. 1.   (a) Schematic representation of a three-dimensional isolated conformally transformed space-time. (b) The geometry of Weyl's stellar compass.

γ through P with κ as tangent that hits $\mathscr{I}^-$ in a point P^-. Let $\tilde{P} \neq P$ be another point in $\mathcal{M}$. Then there will be a unique null-ray $\tilde{\gamma}$ through $\tilde{P}$ that hits $\mathscr{I}^-$ in $\tilde{P}^-$ and whose tangent in $\tilde{P}^-$ is 'parallel' to that of γ in P^- ($\tilde{P}^-$ has the same angular coordinates as P^-). In this sense one might think of $\mathscr{I}^-$ as mathematically representing the celestial sphere and the trajectories γ and $\tilde{\gamma}$ as being two light rays originating from one and the same star if the star (or extragalactic radio source, QSO) is sufficiently remote to reveal no significant parallax or proper motion. Calling tangent vectors to γ parallel to those of $\tilde{\gamma}$ we see that the introduction of $\mathscr{I}^-$ (the celestial sphere) allows to define a global teleparallelism of tangent vectors to light rays from one and the same fixed star. This construction might be viewed as a modern version of Weyl's stellar compass (Weyl, 1970) and has been first introduced by Schmidt, Ehlers and co-workers (Schmidt, 1979; Bonatti-Gonzales, 1981).

Now, for calculating photon paths and astrometric effects in a weak gravitational field the astrophysicist likes to have some 'flat space + forces' concept available so that euclidean vector space notation, etc., can be used. To discuss such a concept we first observe that for directions not directly towards the giant gas planets ($\Delta\varphi|_{R_{2}} \sim 16 \times 10^{-3}{''}$; $\Delta\varphi_{R_n} \sim 6 \times 10^{-3}{''}$) and for accuracies of about $10^{-3}{''}$ we only have to deal with the spherical part of the solar gravitational field in the post-Newtonian approximation: magnitudes for the light deflection due to the post-post-Newtonian contribution of the solar Schwarzschild field, from the Sun's gravitational quadrupole moment and angular momentum are about $11 \times 10^{-6}{''}$, $\lesssim 0.2 \times 10^{-6}{''}$, and $0.7 \times 10^{-6}{''}$, respectively (Epstein and Shapiro, 1980) in Einstein's theory of gravitation. Now, for photons, the post-Newtonian–Einstein theory agrees with the linearized

Einstein theory and having selected a certain gauge (coordinate condition) for the gravitational field the whole theory – field equations plus gauge condition – can be formulated as a geometric, i.e., coordinate independent theory over flat Minkowski space. Such a theory essentially represents a bimetric theory $(\mathcal{M}, \eta, g)$ where η is the flat background metric allowing for the use of euclidean vector space notation and g is the metric describing the rates of atomic clocks.

Let t (world time) be the time an asymptotic clock at rest, e.g., w.r.t. the barycenter of the solar system would measure. Then t causes a certain slicing of space-time into space and time, a necessary ingredient if we want to talk about forces. If we select harmonic (or isotropic) coordinates the post-Newtonian equation of motion for a light ray with incident $(t = t_0,\ \mathbf{x} = \mathbf{x}_0)$ direction $\hat{n}$ becomes, in Einstein's theory,

$$\ddot{\mathbf{x}} = 2m[-\mathbf{x}/r^3 + 2\hat{n}(\mathbf{x}\cdot\hat{n})/r^3]\ ; \tag{1}$$

that readily gives

$$\dot{\mathbf{x}} = \hat{n} - 2m\left[\frac{\hat{n}}{r} + \frac{\mathbf{d}}{d^2}\left(\frac{\mathbf{x}\cdot\hat{n}}{r} - \frac{\mathbf{x}_0\cdot\hat{n}}{r_0}\right)\right] \simeq$$

$$\simeq \left(1 - \frac{2m}{r}\right)\hat{n} - \frac{2m\mathbf{d}}{d^2}\,(\cos\chi + 1)\,, \qquad (r_0 \gg r_\oplus) \tag{2}$$

where χ is the 'unperturbed' angle between the directions towards Sun and source S (see Figure 2), $\mathbf{d}$ is a vector connecting the barycenter of the solar system (origin of

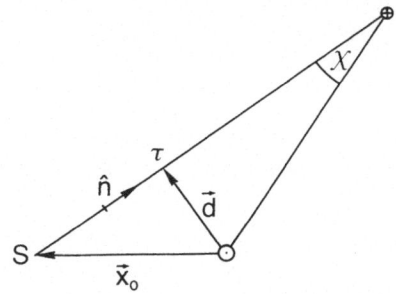

Fig. 2.   Notations used for describing light deflection.

Schwarzschild field) with the point of closest approach to the photon trajectory τ and $m = GM_\odot/c^2 = 1.48$ km. If one is interested in the angle between two light rays τ and τ' with incident directions $\hat{n}$ and $\hat{n}'$ that an observer with 4-velocity u measures we first have to project the two tangent 4-vectors $(c = 1)$ $\kappa^\mu = dx^\mu/dt = (1, \dot{\mathbf{x}})$ and $\kappa' = (1, \dot{\mathbf{x}}')$ into the spacelike hypersurface of the observer and then take the scalar product w.r.t. $g_{\mu\nu}$: in doing so we find that

$$\bar{\kappa}^\mu \equiv \kappa^\lambda p_\lambda^\mu\ ; \qquad \bar{\kappa}'^\mu \equiv \kappa'^\lambda p_\lambda^\mu\ ; \qquad p_\lambda^\mu = \delta_\lambda^\mu + u_\lambda u^\mu\ ;$$

$$\cos\varphi = \frac{\langle \bar{\kappa}, \bar{\kappa}' \rangle}{|\bar{\kappa}|\,|\bar{\kappa}'|}\ . \tag{3}$$

For an observer moving with barycentric velocity $\boldsymbol{\beta}_{\oplus}$ this formula yields (see also Will, 1981; or Brumberg, 1981)

$$\cos \varphi = \hat{n} \cdot \hat{n}' + (\hat{n} \cdot \hat{n} - 1)[\boldsymbol{\beta}_{\oplus}(\hat{n} + \hat{n}') + (\hat{n} \cdot \boldsymbol{\beta}_{\oplus})^2 +$$

$$+ (\hat{n}' \cdot \boldsymbol{\beta}_{\oplus})^2 + (\hat{n} \cdot \boldsymbol{\beta}_{\oplus})(\hat{n}' \cdot \boldsymbol{\beta}_{\oplus}) - \beta_{\oplus}^2] -$$

$$- 2 \left\{ \left(\frac{m}{d} \right) \left(\frac{\mathbf{d} \cdot \hat{n}}{d} \right) \left(\frac{\mathbf{x}_{\oplus} \cdot \hat{n}}{r_{\oplus}} - \frac{\mathbf{x}_0 \cdot \hat{n}}{r_0} \right) + \left(\frac{m}{d'} \right) \left(\frac{\mathbf{d}' \cdot \hat{n}}{d'} \right) \left(\frac{\mathbf{x}_{\oplus} \cdot \hat{n}'}{r_{\oplus}} - \frac{\mathbf{x}_0' \cdot \hat{n}'}{r_0'} \right) \right\}. \quad (4)$$

If one is interested in the astronomical coordinates of a stellar image, the spatial axes defining α and δ have to be introduced into the theory explicitly. In the isotropic gauge the spatial triad of orthonormal vectors $\lambda_{(i)}$ ($i = 1, 2, 3$) is simply given by

$$\lambda_{(i)} = \left(1 - \frac{m}{r} \right) \hat{\lambda}_{(i)}, \qquad (i = 1, 2, 3), \quad (5)$$

where $\hat{\lambda}_{(i)}$ are the astronomical unit vectors in the absence of the solar gravitational field. Neglecting effects from aberration we get

$$\left\langle \lambda_{(i)}, \frac{\overline{\kappa}}{|\kappa|} \right\rangle = \left(1 + \frac{2m}{r} \right) \hat{\lambda}_{(i)} \cdot \dot{\mathbf{x}} \quad (6)$$

or ($r_0 \gg r_{\oplus}$)

$$\overline{\hat{m}} = \hat{m} + \frac{2m \mathbf{d} \cdot \hat{\lambda}}{d^2} (\cos \chi + 1), \quad (7)$$

with

$$\overline{m}^x = - \left\langle \lambda_{(1)}, \frac{\overline{\kappa}}{|\kappa|} \right\rangle = \cos \overline{\delta} \cos \overline{\alpha}, \qquad \overline{m}^y = \cos \overline{\delta} \sin \overline{\alpha},$$

$$\overline{m}^z = \sin \overline{\delta}; \quad \text{etc.}$$

This immediately leads to

$$\tan (\overline{\alpha} - \alpha) \simeq \Delta \alpha = \frac{\delta \Theta}{\sin \chi} \sec \delta \cos \delta_{\odot} \sin (\alpha - \alpha_{\odot}), \quad (8a)$$

$$\tan (\overline{\delta} - \delta) \simeq \Delta \delta = \frac{\delta \Theta}{\sin \chi} \{ \sin \delta \cos \delta_{\odot} \cos (\alpha - \alpha_{\odot}) - \sin \delta_{\odot} \cos \delta \}; \quad (8b)$$

with

$$\frac{\delta \Theta}{\sin \chi} = \frac{2m}{r_{\oplus} \sin^2 \chi} (1 + \cos \chi); \qquad \frac{2m}{r_{\oplus}} \simeq 4.07 \times 10^{-3 ''},$$

$$\cos \chi = \sin \delta \sin \delta_{\odot} + \cos \delta \cos \delta_{\odot} \cos (\alpha - \alpha_{\odot}).$$

Similarly, defining axes comoving with the Earth, we find that, for the aberration alone
a relation similar to (7)

$$\bar{m} = \frac{\hat{m} + \boldsymbol{\beta}(\gamma + \sigma)}{\gamma(1 + \boldsymbol{\beta}\hat{n})} , \qquad \gamma \equiv (1 - \beta^2)^{-1/2} ,$$

$$\sigma \equiv \frac{(\gamma - 1)}{\beta} \cos \Theta ; \tag{9}$$

and (see also Stumpff, 1979)

$$\tan(\bar{\alpha} - \alpha) = - \frac{\sec \delta(\gamma + \sigma)\beta_-}{1 + \sec \delta \beta_+ (\gamma + \sigma)}$$

$$= - \beta_- \sec \delta[1 + \tfrac{1}{2}\beta \cos \Theta - \beta_+ \sec \delta] + 0(\beta^3) , \tag{10a}$$

$$\tan(\bar{\delta} - \delta) = \frac{(\gamma + \sigma)\beta_z \cos \delta - (\gamma + \sigma)\xi \sin \delta}{1 + (\gamma + \sigma)\beta_z \sin \delta + (\gamma + \sigma)\xi \cos \delta}$$

$$= (\beta_z \cos \delta - \beta_+ \sin \delta)(1 + \tfrac{1}{2}\beta \cos \Theta - \beta_z \sin \delta - \beta_+ \cos \delta) -$$

$$- \tan \delta \beta_-^2 + 0(\beta^3) ; \tag{10b}$$

with

$$\xi \equiv \beta_+ - \beta_- \tan[\tfrac{1}{2}(\bar{\alpha} - \alpha)] ,$$

$$\beta_+ \equiv \beta_x \cos \alpha + \beta_y \sin \alpha ,$$

$$\beta_- \equiv \beta_x \sin \alpha - \beta_y \cos \alpha .$$

Thus, according to aberration the apex angle Θ of a stellar image is shifted by an amount
equal to

$$\Delta\Theta = \beta \sin \Theta - \tfrac{1}{4}\beta^2 \sin 2\Theta + \tfrac{1}{6}\beta^3 \sin \Theta (\sin^2 \Theta + 3 \cos^2 \Theta) + 0(\beta^4) . \tag{11}$$

For the barycentric motion of the Earth ($\beta_\oplus^{bc} \simeq 10^{-4}$) the amplitude of the β^2-term is
about half a milliarcsecond. The question remains how the relations for the gravitational
light deflection (4) and (8) to post-Newtonian order are effected by a gauge transformation.

Since we are only interested in the spherical part of the solar gravitational field we
do not want to change the meaning of angle variables and also keep the world time t;
then the most general gauge transformation is a transformation of the radial variable
r of the form

$$r \to \rho = r + \eta(\rho)m \tag{12}$$

(Brumberg and Finkel'shtein, 1979). Now the 'forces' a photon experiences look
different. Instead of Equations (1) and (2) one obtains

$$\ddot{\mathbf{x}} = 2m[(-1 - \eta/2 + \rho\eta'/2)\mathbf{x}/\rho^3 + (2 - \eta + \rho\eta')\hat{n}(\mathbf{x} \cdot \hat{n})/\rho^3 +$$

$$+ \tfrac{3}{2}(\eta - \rho\eta' + \tfrac{1}{3}\eta''\rho^2)\mathbf{x}/\rho^5(\mathbf{x} \cdot \hat{n})^2] \tag{13}$$

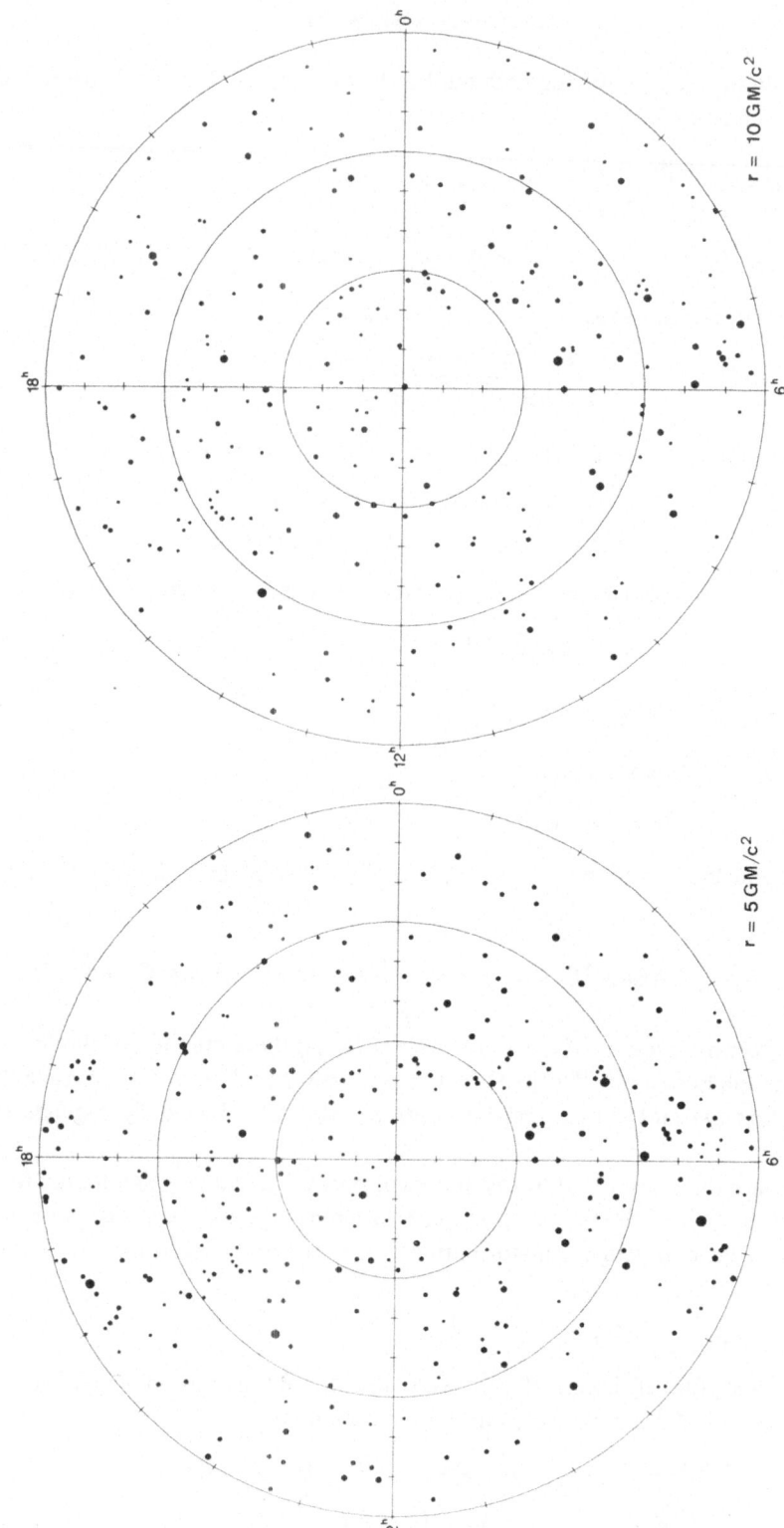

Fig. 3. The stellar sky as seen from the vicinity of a strongly gravitating source.

and

$$\dot{\mathbf{x}} = \hat{n} - 2m \left[\frac{\hat{n}}{\rho} + \frac{\mathbf{d}}{d^2} \left(\frac{\mathbf{x} \cdot \hat{n}}{\rho} - \frac{\mathbf{x}_0 \cdot \hat{n})}{\rho_0} \right) + \frac{\eta}{2} \frac{\mathbf{x} \times (\mathbf{x}_0 \times \hat{n})}{\rho^3} - \frac{\eta'}{2} \frac{(\mathbf{x} \cdot \hat{n})}{\rho^2} \mathbf{x} \right].$$

(14)

However, since relations (4) and (8) were derived from scalar quantities they are unaffected by coordinate transformations.

Now, the small light deflection angle that HIPPARCOS or the space telescope will be able to measure is hard to visualize. The angle of $10^{-3\,\prime\prime}$ correspond to the apparent diameter of a German two-penny coin ($\varnothing \simeq 2$ cm) as seen from a distance of 4000 km! To visualize the effects from light deflection we have plotted the stellar sky as seen from the vicinity of a strongly gravitating source (black hole) (see Figure 3). In Figure 3 the observer's 'radial distance' towards the gravitating source is $10\,GM/c^2$ (Figure 3a) and $5\,GM/c^2$ (Figure 3b).

Acknowledgements

The present study has been prepared as part of the research program of the Sonderforschungsbereich 78 (SFB 78) Satellitengeodäsie of the Technical University of Munich. Support from the Deutsche Forschungsgemeinschaft (DFG) is gratefully acknowledged.

References

Bonatti-Gonzales, J.: 1981, Diploma Thesis, University of Munich, F.R.G.

Brumberg, V. A.: 1981, in E. Gaposchkin and Kolaczek (eds.), *Reference Coordinate Systems for Earth Dynamics*, D. Reidel Publ. Co., Dordrecht, Holland.

Brumberg, V. A. and Finkel'shtein, A.: 1979, *Soviet Phys. JETP* **49**, 749.

Epstein, R. and Shapiro, I.: 1980, *Phys. Rev.* **D22**, 2947.

Schmidt, B.: 1979, in J. Ehlers (ed.), *Isolated Gravitating Systems in General Relativity*, North-Holland, New York.

Stumpff, P.: 1979, *Astron. Astrophys.* **78**, 229.

Weyl, H.: 1970, *Raum, Zeit, Materie,* Springer Verlag, Berlin.

Will, C.: 1981, *Theory and Experiment in Gravitational Physics*, Cambridge Univ. Press, Cambridge.

FRIEDRICH WILHELM BESSEL
1784, JULY 22–1846, MARCH 17*

P. VAN DE KAMP

Sterrenkundig Instituut, Universiteit van Amsterdam, Amsterdam, The Netherlands

Bessel had a very early inclination to calculation; he decided to become a businessman with a specialty as ship-broker. He studied navigation with enthusiasm, including the work of Lalande and of Olbers, as well as higher mathematics. His famed results on astronomy led to his appointment in 1810 as director of the then new observatory in Königsberg, planned by King Friedrich Wilhelm III of Prussia. With initially limited means, he obtained valuable results, both to instrumentation and theory. Stellar catalogues included the observations of James Bradley (Fundamenta Astronomiae, Königsberg, 1818). Further catalogues of stars were worked on for the Berlin Academy. Bessel worked on one of the established 'zones', about one degree wide in declination, cataloging accurate positions of stars. He noted that the tails of comets are affected by both, the comet and the Sun.

The Fraunhofer heliometer acquired in 1829 was tested and used by Bessel. This led to important astrometric results. Both the heliometer and meridian approach proved to be of special importance. Bessel's practical and spherical astronomy as well as his work together with Bayer resulted in significant reinformation on the shape and size of the Earth.

One of the first accurate determinations of the annual stellar parallax was made in 1838 by Bessel for the double star 61 Cygni. Because of its large annual proper motion of 5.22″, the system was judged to be nearby. Bessel used the differential method, measuring 61 Cygni on a background of reference stars at a small angular separation. The instrument was the heliometer, a double image micrometer with two halves sliding by each other. This accurate method yielded a value of 0.35″ for the parallax, which is now improved by the 'modern' value of 0.29″.

The differential method yielded the first success in this field by finding the parallax relative to the reference stars and eliminates to a great extend the spherical effects of the observations (precession, refraction, instrumental effects).

61 Cygni is actually a 'wide' double star, with the two components separated by less than one minute of arc. The two components of 61 Cyg have about the same mass, absolute magnitudes 7.5 and 8.3, and spectral types K5 and K7.

Gradually, the visual method was replaced by the photographic method, the accuracy going well below 0.005″.

---

* Communication prepared for the International Conference on 'Astrometric Binaries', held on June 13–15, 1984, at the Remeis-Sternwarte Bamberg, Germany, to commemorate the 200th anniversary of the birth of Friedrich Wilhelm Bessel (1784–1846).

Unfortunately, illness prevented the author from presenting this communication in person.

Astrophysics and Space Science **110** (1985) 103–104. 0004–640X/85.15

In 1844 Bessel announced the perturbations in the proper motions of the bright nearby stars, Sirius and Procyon. The companions were not seen until 1862 and 1896.

The unseen planets Neptune (1846) and Pluto (1930) were not discovered until after Bessel's death.

The great portion of the found unseen companions has more than 10% of the mass of our Sun, close to the lower limit for a normal star. Some other unseen companions have smaller masses – i.e., the dark dwarfs with minute luminosity, which are called 'dark dwarfs' or 'star planets'. For one star, Barnard's star, at a distance of 6 light years, there seem to be two planets with masses of 0.7 and 0.5 times Jupiter's mass, and periods or revolution of 12 and 20 years. This would seem the first indication of a planetary system outside our own planetary system, of which the two heaviest planets have now been found.

CURRENT WORK ON BINARY AND MULTIPLE STARS*

MARIO G. FRACASTORO

President, IAU Commission No. 26

Osservatorio Astronomico di Torino, Pino Torinese, Italia

(Received 10 July, 1984)

Abstract. Following the kind invitation of Prof. Kopal, I am going to refer at this Conference on current work carried out by the IAU Commission 26, after having circulated a letter to all members to collect recent information at various places. I am grateful to those who replied in time for this report, which will also contain some personal considerations.

1. Name and Topics

Commission 26 has recently changed its name to 'Double and Multiple Stars'. The new denomination, warmly promoted by Dr Dommanget, in fact represents more adequately the actual research field of the Commission. Furthermore, the initiative appeared to me as a good start for extending its interest beyond the traditional aspects (morphology and statistics) and plunging into problems connected with the origin, evolution, mass spectrum and dynamical aspects of these small associations of stars.

My personal hope is that in the near future Commission 26 may establish a fruitful partnership with other commissions. On this point, I should refer that there has been an exchange of correspondence with Prof. Mauri Valtonen, from Helsinki, who asked Commission 26 to co-sponsor a meeting dedicated to 'Gravitational Few-Body Problem' (many-body effects in the solar system, galactic dynamics, star clusters, multiple star systems). The planned meeting is supported by Profs. J. Kovalevsky and V. Szebehely. Future developments of this initiative will be very welcome.

2. Observations

It is well known that the number of expert observers, mainly dedicated to the discovery and systematic observations of binary and/or multiple stars is rather scarce, in spite of the urgent need of new and better data. Perhaps, the most popular instrumentation for accomplishing this type of research is still based on classical tools: a long focus refractor and the filar micrometer. Most of these instruments were inaugurated at the end of the 19th century or at the beginning of the present one.

However, improved techniques have been developed and special apparatuses are at work in several observatories. I do not want to invade the topic which has been reserved

---

* Communication presented at the International Conference on 'Astrometric Binaries', held on 13–15 June, 1984, at the Remeis-Sternwarte Bamberg, Germany, to commemorate the 200th anniversary of the birth of Friedrich Wilhelm Bessel (1784–1846).

to other speakers and I will only quote the areal scanners active at the Lowell Observatory, in Vienna and other places, as well as the speckle interferometers at Atlanta, Georgia, and in France.

The goal of these auxiliary instruments is not only the improvement of the quality of the measurements: namely, to have better ρ's and ϑ's, as well as better Δm's, but also to yield objective data, avoiding the risk of personal errors and consequently a lack of coherence among the various observers. Perhaps, there is a third advantage: that of offering young neophytes a modern and more attractive tool, even if the mise-à-point of these apparatuses is much more critical than that of the classic filar micrometers.

Systematic patrols, run by means of visual observations, are in course in several countries. Thanks to a kind communication received from Prof. P. Couteau, I may quote the most up to date status of research in France.

P. Couteau and P. Muller are continuing with their systematic patrol of the sky, for $\delta > 52°$ (Muller) and $+17° < \delta < +52°$ (Couteau), using the two refractors of 76 and 50 cm aperture at Nice. At present (1984) Couteau has patroled 102 314 stars out of the total amount of 169 243 BD stars contained in the zone, discovering 2200 new couples. Muller has already patrolled 35 000 stars, discovering 650 new couples. Thanks to their assiduous work, the survey of the Northern Hemisphere is now as good as that of the Southern one (Rossiter, van den Bos, Innes). We may compare the situation as it was in 1961 (according to the *Index Catalogue* data), and as it is in 1984:

Separation	Hemisphere		
	Southern	Northern	
	1961	1961	1984
$s < 0\rlap{.}''25$	946	661	1066
$s < 0.50$	3385	2152	3119
$s < 1.0$	6260	3777	5454

It should be pointed out that data concerning the southern hemisphere are not updated.

To remain in France, at CERGA, Alpes Maritimes, Bonneau, Foy, and Koechlin are using speckle interferometry techniques with 'large' telescopes, in the wake of Dr Labeyrie. A further list of measurements is expected since the one published in 1980.

Finally, it is a pleasure for me to say that Dr Baize is still working eagerly in his private observatory in Normandie and publishes about 15 orbits per year.

Other orbits have recently been published by R. R. de Freitas Murão in Rio de Janeiro. He is also an active observer.

May I be allowed to add that in Torino G. Massone, after having started systematic observations with the old 30-cm Merz refractor, is now continuing his plan with the 42-cm refractor, whose lens was recently reconditioned and has now a new mechanical mounting. Astrometric observations of wide couples (separation and parallaxes) are also on the schedule with the 41-inch astrometric telescope (Pannunzio, Massone, and Morbidelli).

3. Catalogues and Statistics

A very important work has been accomplished at the U.S. Naval Observatory by Ch. Worley with his *Washington Double Star Catalog*. He has completely edited, corrected, added to and revised the previous Index and Observations Catalogue, which is now updated with all published observations up to 1 January 1984 and should be available to all users, having been transmitted to the various Data Centers, for instance to that of Strasbourg (Prof. Jaschek).

Worley and Heintz have just published their *4th Catalog of Orbits of Visual Binary Stars* (Publ. USNO, Vol. XXIV, Part VII).

Before coming to statistics, let me start with a personal but not completely obvious remark. My opinion is that it would be opportune, if not necessary, to tune the various criteria which have been adopted by different Authors to establish the percentage of doubles or multiples among stars. In my opinion, α *Canis Majoris* is a star, not only for historical reasons, but also because it occupies a single entry, or box, in all Catalogues (*Bright Stars* of Yale, *Nearby Stars* by Gliese, and so on, going back to Hipparchos and Ptolemaeus). It has been shown by F. W. Bessel, and actually observed several years later, the Sirius is actually a binary star (α CMa A, α CMa B). Hypothetically, future observations might prove the presence of a third body, which will be named α CMa C.

Given that, we may proceed and say that, out of 4548 entries ($RA < 12^h$), the *Bright Star Catalogue* of Yale contains 1073 visual pairs: namely 23.6%, if we consider the number of entries, but 38.2%, if we compute all single objects. In other words, out of 1000 boxes (which may be also interpreted as space boxes) 764 are occupied by single stars and 236 by two or more stars. Stars which are considered up to now as single, may become doubles; double stars may become triples and so on. However, the number of boxes will always remain the same.

Let us now come to statistics. Up to the 9th magnitude, we have a precise idea about the percentage of duplicity among stars. Plotting the total number N of stars and the number D of known doubles, in terms of the apparent magnitude m, the two lines ($\log N$, m and $\log D$, m) run parallel. For $m = 10$ or greater, there is a sudden divergence, obviously due to severe incompleteness of observations. The impact of duplicity and multiplicity among stars should obviously consider all kinds of binaries. I have shown several years ago that about 6 percent of all stars brighter than 2nd magnitude are eclipsing binaries. This fraction falls to 1% already for $m = 5$. Of course, we should multiply these percentages by csc i, where i is the limiting angle of inclination for having an observable eclipse. Unfortunately, this angle depends strongly on the type of eclipsing binary and becomes very small for detached systems.

Coming now to recent statistical research, it appears to me worthwhile to quote a paper published by Poveda *et al.* (1982). Working on the previous Index, they elaborated a *Filtered Index of Double Stars*, with the aim of handling a more suitable set for statistical purposes, after the elimination of optical pairs. The selection criterion is the following: a secondary component of magnitude m_2 is only accepted as a physical member of the neighbouring star when the probability of being at distance s is smaller than 1%, namely

$$\pi s^2 N(m_2)_{l,\,b} < 0.01 ,$$

where $N(m_2)_{l,\,b}$ is the average number of stars of magnitude m_2 per unit area in a region of the sky having galactic coordinates l and b.

As a consequence of this filtering process, 18 618 boxes have been eliminated and 51 201 have been retained, being considered as physical pairs, at least for statistical purposes.

For $2.5 < m < 6$, Poveda, Allen, and Parrao give the relation

$$Q(m) + Q'(m) = 0.675 \times 10^{-0.095}\, m ,$$

where $Q(m)$ and $Q'(m)$ indicate the probability that a star, having been examined for its possible duplicity or multiplicity, has actually been registered as a binary or a multiple, respectively.

Extrapolating to $m = -1.3$, the result is that the probability to be double tends to 1. And Sirius is indeed double!

An analysis of the *Bright Star Catalogue* has been made recently (Fracastoro, 1983), concerning its content of visual pairs, regardless their physical link. In particular, the Δm's and separations s have been examined. It results that:

(i) plotting the average $\overline{\Delta m}$ in terms of s, the former increase up to very large values of s $(16'')$;

(ii) the observed number of pairs exceeds that foreseen by the Poisson formula even for large separations. Both facts may suggest that low mass companions are revolving on large orbits, or are even pushed to hyperbolic ones.

4. Mass-Luminosity Relation and H–R Diagram

As recently stated by Dr Heintz at the 111 IAU Symposium in Como last May, masses are the final result of a difficult itinerary of observational data. No wonder if the reliable data are only a small fraction of the available material.

Out of 850 pairs which have orbits (only 1% of all known binaries), not more than 40 or 50 are good enough for mass determinations. The same can be said for parallaxes and radial velocities. Finally, a severe selection results in favour of fractional masses ranging between 0.4 and 0.5 (q close to 1). However, the incompleteness of data increases rapidly when Δm exceeds 2.

Rakos and Sinachopoulos (preprint) at Vienna have employed 147 of the best known binary stars, with a total of 294 Main-Sequence components, for a parallax independent calibration of the Main Sequence, in the interval $-0.15 < B - V < +0.80$. The $(M_v, B - V)$ diagram is found to be equal to that obtained from composite cluster diagrams, for ages between 8.35 and 8.43×10^8 yr, very hot stars excepted. A new estimation of the luminosity distribution is presented for multiple systems. This distribution fits in with van Rhijn's as far as the mass ratio q is ≥ 0.5. For $q < 0.5$, the number of close companions diminishes rapidly.

It is well known that the most frequently observed mass ratio q among binaries is close to 1. However, lower values are frequently found, and the reality of a secondary peak for $q = 0.2$ is debated. The question of substellar companions is also well known. One might ask how small q can be for the star still being considered as a binary, or whether masses below the lowest theoretical limit (0.085 $\mathfrak{M}_\odot$) are really observed. Certainly, we are still very far from revealing Earthlike bodies and consequently from solving by this procedure the great problem of finding life on other worlds. However, from a merely astronomical point of view, it appears that even the question of existence and frequence of substellar masses, say 'Jupiters', as secondary components of pairs, has a basic importance. Instruments specially designed for this purpose may be key for getting a reliable answer to the problem in the near future. Alas, even in this case, things cannot be speeded up, because gravitational displacements are probably of the order of 10–20 yr!

5. Origin and Dynamics of Multiple Systems

Fragmentation is the most accepted mechanism for explaining the origin of multiple systems. A decisive parameter is the angular momentum, which appears to be distributed at random in space. Fluctuations of density inside the proto-cloud can be due to turbulence or to chemical processes (for instance, formation of H_2), or to the presence of grains, their abundance being given in terms of the number of hydrogen atoms, n_g/n_H.

A typical proto-cloud is assumed to have the following structure:

$$\mathfrak{M} = 10^3 \, \mathfrak{M}_\odot = 2.7 \times 10^{36} \, \text{g} \,,$$

$$n_H = 4 \times 10^3 \, \text{cm}^{-3} = 7 \times 10^{-21} \, \text{g cm}^{-2} \,,$$

$$2R = 3 \, \text{parsec} = 6 \times 10^{18} \, \text{cm} \,,$$

$$n_g/n_H = 10^{-12} \quad \text{and consequently} \quad \mathfrak{M}_g = 10\mathfrak{M}_\odot \,.$$

The diameter of each grain results to be $\sim 10^{-5}$ cm. This is an average value. Nothing is said about a possible distribution of diameters and masses among the grains, according to a 'law' of the type

$$N \propto a\mathfrak{M}^{-b}$$

(Fracastoro, 1969), with $b = 0.8$–0.9. In this case, a total mass of grains $\mathfrak{M}_g = 10 \, \mathfrak{M}_\odot$ could include 'grains' of significant dimensions, whose mass would change very considerably the conditions for condensation, and in particular the minimum mass required for it (Jeans limit), increasing the agreement between the theoretical and the observed distribution of mass among stars.

In any case, the discovery of faint members in multiple systems may improve the observed mass spectrum and be a valid test for the fragmentation theories.

References

Fracastoro, M. G.: 1969, *Mem. Soc. Astron. Ital.* **40**, 309.
Fracastoro, M. G.: 1983, *Rend. Acc. Lincei* **73**, 226.
Poveda, A., Allen, C., and Parrao, L.: 1982, *Astrophys. J.* **258**, 589.

MODERN METHODS IN THE ASTROMETRY OF
DOUBLE STARS*

LAURENCE W. FREDRICK

Leander McCormick Observatory, University of Virginia, Charlottesville, U.S.A.

(Received 1 June, 1984)

Abstract. The techniques for studying double stars continue to evolve in a predictable way. The most recent major breakthroughs have been the area scanner and speckle interferometry. On the horizon looms the application of large format CCDs which will replace the photographic plate for many astrometric problems.

To my good fortune, this past spring the Lowell Observatory Bulletin No. 167, Vol. IX, No. 1 appeared. This volume, *Current Techniques in Double and Multiple Star Research* (IAU Colloquium No. 62) edited by R. Harrington and O. Franz covers the subject quite well. With this publication as required reading, a brief summary is all that is required. When this publication is referred to in the following it will be abbreviated simply as (1983, LOB) and will not be cited at the conclusion of the paper.

Taking a broad view of the topic, we must recall the purpose of studying double stars. It seems that they have been put in the order of nature to help us unravel the mystery of stellar evolution. Thus, our efforts should be directed to those systems that will give us masses and luminosities and to those methods and studies that will advance this direction. There are other uses for binary stars, but for now we will overlook them.

The determination of masses call for advances not just in double star observing, but in parallaxes as well. And here we do not restrict ourselves to the traditional parallax measurements, but those that can be obtained by any means, e.g., knowing the orbital elements and the radial velocities of the components we can convert the angular separation of the pair into its parallax.

The history of the various techniques used in double star research is more evolutionary than revolutionary. There have been a few landmark developments, e.g., Rakos' area scanner and Labeyrie's speckle interferometry, but in general known techniques have been gradually improved or have had to wait for certain developments, for example high speed electronics so that the lunar occultation method could be applied.

At present we are certainly in an evolutionary stage with almost all available techniques evolving more or less rapidly. The present era is similar to the evolution of the calculating machine from the old hand-cranked models we used as students to the modern pocket calculator of almost infinitely more power and speed. Alas, many of us seem to be forgetting how to use logarithms.

---

* Communication presented at the International Conference on 'Astrometric Binaries', held on 13–15 June, 1984, at the Remeis-Sternwarte Bamberg, Germany, to commemorate the 200th anniversary of the birth of Friedrich Wilhelm Bessel (1784–1846).

Astrophysics and Space Science **110** (1985) 111–115. 0004–640X/85.15
© 1985 *by D. Reidel Publishing Company.*

Perhaps the major current advance in binary star astrometry is the general application of microprocessor based devices, including those in various computers. They are used to control our telescopes, control our measuring machines, and collect data from both. The modern application of analytical techniques would be impossible without the computer.

Modern methods date back to Fraunhofer (Ambronn, 1899) and his screw micrometer which with a few changes became known as the filar (or bifilar) micrometer. I know of at least one budding astronomer who turned to physics because of this instrument. He removed the 'cobwebs' with his index finger so he could have a better view. The bifilar micrometer remained almost unchanged until very recent times. Rakos' (1965) area scanner is a revolutionary advance (with some limitations) in this area and we will say more about this later.

Fraunhofer also constructed fine optics, among them: lenses for refracting telescopes and heliometers. The latter instrument in the hands of Bessel enabled Bessel to measure and publish the first trigonometric parallax (see contributions in this volume by Prof. Geyer and Prof. Strohmeier).

The photographic plate was first applied to double stars by Bond in 1857 and to parallaxes by Pritchard in 1886. Its serious application however developed in two directions with Hertzsprung (1920) and Schlesinger (1910). The only changes in either technique has been a slow trend toward automation and, of course, a better and wider choice of emulsions. The real advances here have been in the measuring machines, control systems, and data reduction and analysis. One can see a revolutionary step developing here, ultimately perhaps, toward large field CCDs.

An ever evolving method for studying binary stars is the photographic method. The Hertzsprung technique applies to well separated pairs and is well known to all of us. While a variation is in use at Torino (Pannunzio and Scarda, 1983, LOB), it is some concern that the method's practitioners are declining. A few astronomers, e.g. Monet (1983, LOB) have applied CCD arrays to this problem with success.

What I have referred o as the Schlesinger technique is still very much in evidence, primarily as a tool for obtaining parallaxes. Because parallax series tend to cover rather long intervals, it is a marvelous tool for uncovering certain types of binaries – astrometric binaries. It mimics in detail the discovery of the first such binary more than 150 years ago by Bessel (1844).

The modern applications trace back to Reuyl (1936) at the McCormick Observatory when he discovered the irregular proper motion of Ross 614. An interesting account of this fortunate event has been given by van de Kamp. The star was later studied in detail by Lippincott (1951) at the Sproul Observatory and resolved by W. Baade (Lippincott, 1955) using the 5-meter Hale telescope.

Even with the great storage capacity of the photographic plate and its permanence, efforts are well underway to replace it. There are two reasons for this: (a) most of the plate is not used even in a multiple field, multiple exposure format; thus most of the silver is wasted, and (b) the cost of plates is escalating faster than the general price index for the world as a whole. The effort to replace the photographic plate is to make use of the photoelectric effect and takes on several variations: one where the stellar images are modulated (Villamediana and Fredrick, 1971; Mertz, 1971; van Altena, 1974; Gatewood, *et al.*, 1980; and Jones, 1981). A variant is to measure positions by interferometric means (Shao and Staelin, 1980). Another will be the use of CCDs.

The modern trend in pure double star observations receives its pressure from two sources, one to resolve very close pairs and a second to make the observations impersonal. We have mentioned the lunar occultation method. However, the revolution here occurred with the work of Rakos on one hand and Labeyrie on the other.

Rakos (1965) developed a rapidly scanning slit photometer and recorded the various scans, first on moving film, then on magnetic tape, and then in a multi-channel scaler. Rotating the photometer head about the optical axis and thus scanning at various position angles yielded the separation and positon angle. A bonus appears in an accurate measure of the Δm in any passband acceptible to the photomultiplier.

Rakos (1983, LOB) has now improved this by replacing the mechanical slit and photomultiplier with a reticon detector. This has the advantage of being a 'solid state' device. The reticon is scanned electronically and each scan is stored on disk. (Note the appearance of the micro-processor here.) Each scan is then fit to an arbitrarily chosen initial scan. Again, doing this at various position angles and behind various filters yields the desired data. This technique, in good seeing, yields information in the range of separations between 0.5 to a few arc-seconds and magnitude differences as large as 3.

For very close pairs various forms of interferometers have been used starting with Michelson (1920) and Anderson (1920). For some reason these have never really taken hold until the revolutionary development of speckle interferometry by Labeyrie. Labeyrie's (1970) speckle interferometry technique makes use of the granularity of the image imposed upon it by the atmosphere. Over the small angle where the seeing is corollated, it can be applied to the theoretical resolving power of the telescope being used. An excellent summary of the technique is given by McAlister (1983, LOB) and he refers you to many earlier papers. This technique covers the range from 20 arc millisec for the largest telescopes to about 1 arc sec and slightly larger. This is a powerful technique because this is the domain of the spectroscopic binaries and hence adds to the number of systems which can be fully determined. Several classic astrometric binaries have been resolved, a few to the embarrassment of the classical photographic technique.

A recent extention of the speckle technique has been developed by Howell *et al.* (1981) where they use infrared detectors and filters. A red star in the presence of a F dwarf is not detectible at visual wavelengths, but may dominate in the near and far infrared. This approach has, for example, solved the long standing problem with Mu Cas.

Other interferometers have been applied to the problem. The pioneering work of Finsen (1951), and Hanbury-Brown (1974) are examples. Of great promise is the work on long baseline interferometers of the amplitude variety. Current efforts in this direction are being pursued by Labyrie (1975), Davis (1981, also 1981, LOB) and probably many others.

We should not overlook the intensity interferometer which is also a long baseline device. It is well described by Hanbury-Brown (1974) and yields the angular diameters of the components, when they have sensible diameters, and the angular separation along with other elements including the sense of orbital motion. When combined with spectroscopic observations one can then determine the parallax, individual masses, etc., for the system including the surface gravity and absolute fluxes. In one famous case – Spica – a semi-major axis of 1.54 arc millisec was measured to 1% accuracy (Herbison-Evans *et al.*, 1971). While there is no activity on this type of instrument at present, it could be revived with a new breed of high quantum efficiency detectors. The desired information is in a second-order effect and, hence, of small amplitude.

In passing, we cannot leave interferometers without just a word about the white-light interferometer used by Finsen (1951). Another such device was designed by Sinton (1954). This is a very effective tool that has found few users.

On occasion, nature provides us with a new technique in a serendipitous fashion, as in the case of lunar occultations. The possibility of detecting binaries by this technique was first discussed by Herschel in 1865 (see White, 1983, LOB), but the application awaited the development of high speed electronics. An excellent review of this technique is given by Evans (1983, LOB) in which he properly laments the unhappy fact that with this technique we are limited to a roughly ten degree band centered upon the ecliptic.

Lastly, I should add the dimension of space (as in outerspace). It is hoped that soon HIPPARCOS and Space Telescope will add greatly to overcoming some of the sticky parts of our science in the study of binary stars. While this is not the direct goal of either satellite, the spin-off will be tremendous. The details of HIPPARCOS are discussed by Dommanget in this volume and of Space Telescope by Fredrick *et al.* (1983, LOB). An astrometric satellite is being proposed and I could envisage a high-speed photometer and telescope combination to extend the band for which occultation observations are possible. Any space station should have a modest aperture telescope. While it might not solve the grand problem of the origin of the universe, it might well be applied to fundamental astronomy, including astrometry.

Acknowledgements

The preparations of this summary could not have been done without the cooperation of my colleagues world-wide. The undue emphasis upon work by our colleagues in the U.S., France, and Australia results from the short time frame for preparation. Material from the Soviet Union and from China has been promised, but had not yet arrived when this was prepared.

References

Ambronn, L.: 1899, *Handbuch der Astronomischen Instrumentenkunde* 2, 521ff. Springer Verlag, Berlin.
Anderson, J. A.: 1920, *Astrophys. J.* **51**, 263.
Bessel, F. W.: 1844, *Astron. Nachr.* **22**, 145.
Davis, J.: 1981, Chatterton Astron. Dept., School of Physics, Univ. of Sydney, Australia (preprint).
Finsen, W. S.: 1951, *Union Obs. Circ.*, No. 112.

Gatewood, G., Breakiron, L., Goebel, R., Kipp, S., Russell, J., and Stein, J.: 1980, *Icarus* **41**, 205.

Hanbury-Brown, R.: 1974, *The Intensity Interferometer*, Taylor and Francis, London.

Herbison-Evans, D., Hanbury-Brown, R., Davis, J., and Allan, L. R.: 1971, *Monthly Notices Roy. Astron. Soc.* **151**, 161.

Hertzsprung, E.: 1920, *Publ. Astrophys. Obs. Potsdam* **24**, No. 75.

Howell, R. R., McCarthy, D. W., and Low, F. J.: 1981, *Astrophys. J.* **251**, L21.

Jones, B. W.: 1981, *Lick Obs. Ann. Report*.

Labeyrie, A.: 1970, *Astron. Astrophys.* **6**, 85.

Labeyrie, A.: 1975, *Astrophys. J.* **196**, L71.

Lippincott, S. L.: 1951, *Astrophys. J.* **55**, 236.

Lippincott, S. L.: 1955, *Astrophys. J.* **60**, 379.

Mertz, L.: 1971, *Optica Acata* **18**, 51.

Michelson, A. A.: 1920, *Astrophys. J.* **51**, 257.

Rakos, K. D.: 1965, *Appl. Opt.* **4**, 1453.

Reuyl, D.: 1936, *Astrophys. J.* **45**, 144.

Schlesinger, F.: 1910, *Astrophys. J.* **32**, 372.

Shao, M. and Staelin, D. H.: 1980, *Appl. Opt.* **19**, 1519.

Sinton, W. M.: 1954, *Astrophys. J.* **59**, 369.

van Altena, W.: 1974, in W. Gliese, C. A. Murray, and R. H. Tucker (eds.), 'New Problems in Astrometry', *IAU Symp.* **61**, 311.

Villamediana, J. F. and Fredrick, L. W.: 1971, *Proc. IAU Colloq.* **11**, 95.

ON THE OBSERVATION OF DOUBLE STARS FOR THE DETERMINATION OF THEIR PROPER MOTIONS*

M. S. ZVEREV

Main Astronomical Observatory, USSR Academy of Sciences, Pulkovo, Leningrad, U.S.S.R.

(Received 11 June, 1984)

Abstract. The state of meridian observations of the DS program is reported. The program contains close double stars whose precise positions can be determined by the visual method only. In some declination zones these stars have been observed at a few observatories of the USSR and in Strasbourg, Santiago de Chile and Beograd. The importance of further DS observations, particularly in the southern sky, is stressed.

The determination of proper motions of stars fainter than 7th magnitude is generally carried out using their photographic positions in catalogues (AGK2, AGK3, *Yale Catalogues, Cape Catalogues*, and some others).

However, in all the above catalogues those close double stars are absent, whose photographed components merge, thus making it impossible to precisely determine their coordinates.

These stars are also not available for meridian observations with photoelectric micrometers. Hence, old epochs of observations of many double stars can be found in the visual zone catalogue AGK1 of the last century only. As a rule these stars are absent from the catalogues of the 20th century. Their proper motion determination appears, therefore, actually impossible.

In the 1920s, when Bergedorf and Bonn observatories began a photographic repetition of the AGK1, Courvoisier (1949) in the Berlin-Babelsberg observatory carried out visual meridian observations of 1668 stars of the northern sky, mainly double stars, whose coordinates could not be determined photographically with the zone astrograph (d = 16 cm, f = 200 cm). The results of these meridian observations are published as a catalogue (Courvoisier, 1949) and as short supplements to each zone of the photographic AGK2.

The program 'DS' ('double stars') of 3000 double stars from 6.0 to 9.0 magnitudes, that might be troublesome to measure on photographic plates, was prepared at the Pulkovo observatory (Zverev and Timoshkova, 1960). That undertaking was connected with the compilation of the AGK3, headed by the late Dr Dieckvoss. The DS were recommended for meridian observations at the 10th General Assembly of the IAU in Moscow, in 1958. However, the majority of meridian instruments were then engaged in other international programs (AGK3, BS, KSZ, SRS, etc.), and the DS list was only

---

* Communication presented at the International Conference on 'Astrometric Binaries', held on 13–15 June, 1984, at the Remeis-Sternwarte Bamberg, Germany, to commemorate the 200th anniversary of the birth of Friedrich Wilhelm Bessel (1784–1846).

observed by Dr P. Lacroute and others at the Strasbourg observatory (1027 DS in the zone $+25°$ to $+75°$ mean epoch is about 1965).

Dr Scott (1967) revised the Pulkovo DS list in connection with the international undertaking for meridian observations of SRS and BS stars in the sixties. He extended the list to the southern sky using the new index catalogue of double stars of the Lick observatory (Jeffers and Van den Bos, 1963; Scott, 1967); Scott's list contains 2292 double stars. Merdian observations of 509 stars in this list with declination from $-47°$ to the South Pole were carried out by the Pulkovo astronomers at the Cerrol-Calan Observatory (Chile) in 1963–1968.

Only in the late seventies, after the recommendation of DS observations was reaffirmed at the 15th General Assembly of the IAU Sydney, six observatories of the U.S.S.R. (Moscow-Sternberg Astronomical Institute, Kazan-Engelhardt Lbservatory, Kiev, Kharkov, Odessa, Tashkent) and Beograd Observatory in Yugoslavia included DS stars in their visual meridian programs. Moscow astronomers extended the DS list with the introduction of another 266 stars of the northern sky, which are of special interest for stellar astronomy. At present, the committment has been nearly completed. The first four of the above U.S.S.R. observatories have obtained 16 observations in each coordinate for 800 DS stars with declination northward of $+30°$. The observations are more scarce in the equatorial zone ($-30°-+30°$) particularly to the south of the equator. In Beograd, DS were also mainly observed in the equatorial zone.

In general, the status of DS observations cannot be considered satisfactory. It particularly concerns the southern hemisphere, where there are practically no observations of DS in the declination zone from $-10°$ to $-47°$. I hope very much that the U.S.S.R. observatories and perhaps some observatories of other countries will be able to carry out observations of DS in the large equatorial zone in the nearest future. The stars from $-20°$ to the South Pole could be observed by the southern observatories (Santiago-Chile, San Juan, Cape Observatory). Of course the problem of an elaboration of the technique for automatic registration of double stars with a separation of the components $2-3''$ and less is very urgent. But, at present, we do not dispose of such a reliable and tested method and that is why it is desirable to have some meridian instruments for visual observations on at least few observatories.

References

Courvoisier, L.: 1949, *Veröff. Berlin-Bab.* **12**, 6.
Jeffers, H. M. and Van den Bos, W. H.: 1963, *Lick Publ.* **21**, 1.
Scott, F. P.: 1967, *Trans. IAU* **XIIa**, 73.
Zverev, M. S. and Timashkova, G. M.: 1960, *Transactions on 14th USSR Astronomical Conference*, Nauka, p. 147.

ON THE DETERMINATION OF DOUBLE STAR ORBITS* **

HEINRICH EICHHORN

Department of Astronomy, University of Florida, Gainesville

(Received 20 August, 1984)

Abstract. It is proposed to improve the convergence of the determination of the elements of the orbits of visual binaries by using not only first, but second-order derivatives in the development of the appropriate equations of condition. Also, some improvements of the Kowalsky–Seeliger method are suggested which improve the accuracy of the orbit used as first approximation.

1. Critique of the Kowalsky–Seeliger Method

The reputation of this method – described in the standard literature (cf. Heintz, 1971) – points out that good results are not likely to be obtained because the relative rectangular observed coordinates (x_v, y_v) of the secondary with respect to the primary are subjected to the condition

$$F_v(x, \alpha) = \alpha x_v^2 + 2\beta x_v y_v + \gamma y_v^2 + 2\delta x_v + 2\varepsilon y_v - 1 = 0 , \tag{1}$$

which involves five of the seven necessary orbital elements as adjustment parameters and does not enforce the area theorem. However, it is clear that the area theorem

$$\dot{x}y - \dot{y}x = c , \quad \text{for all pairs } x, y , \tag{2}$$

can be usefully employed as a condition equation only by utilizing also the epochs t_v at which the observations were made. Superficially, it appears that the velocity components in Equation (2) make it impossible to enforce the area theorem from quantities which are directly accessible to observation. It is, however, clear that the velocity components are obtained by differentiating the functions $x(t)$ and $y(t)$ with respect to the time. We know, however, that because of the transcedental nature of Kepler's equation, straight differentiation only leads to

$$\dot{x} = \dot{x}(x, y, t) ; \quad \dot{y} = \dot{y}(x, y, t) . \tag{3}$$

These equations, used in Equation (2), should yield the required equation of condition which contains as arguments of the functions which it connects, observations (or rather observable quantities) and adjustment parameters exclusively. In any case, Equation (1) alone is a straightforward equation of condition, even though it does not make use of the epochs at which the observations were made.

There are well-known relationships between the coefficients $\alpha^T = (\alpha, \beta, \gamma, \delta, \varepsilon)$ in Equation (1) and the orbital elements $i, e, a, \Omega,$ and ω. Equation (1) has the advantage

---

* Communication presented at the International Conference on 'Astrometric Binaries', held on 13–15 June, 1984, at the Remeis-Sternwarte Bamberg, Germany, to commemorate the 200th anniversary of the birth of Friedrich Wilhelm Bessel (1784–1846).
** Communication from the Dept. of Astronomy, University of Florida, Gainesville, No. 77.

Astrophysics and Space Science **110** (1985) 119–124. 0004–640X/85.15
© 1985 *by D. Reidel Publishing Company.*

that it appears to be simple – specifically, linear in the adjustment parameters. Its application to the determination of orbits, so far, consisted in regarding the right-hand sides of the equations which result from inserting a pair (x_v, y_v) of observed rectangular coordinates into Equation (1), as errors with a univariate gaussian distribution (normally distributed errors) and performing a least-squares adjustment which is linear in the adjustment parameters. This approach, while it has the advantage that approximation values for the adjustment parameters need not be available at the outset, fails to take into account two facts.

(a) It is not the right-hand sides of the condition equations which are to be considered as normally distributed errors, but rather the observations (x_v, y_v) or, in the case that polar coordinates were observed, ρ_v and P_v.

The condition equations (1) thus contain more than one observation each. Since the observations x_v, y_v (or ρ_v, P_v, as the case may be) occur in the condition equations nonlinearly, the matrix $\mathbf{X} = (\partial F(x, a)/\partial x)_{x = x_0, \, \alpha = \alpha_0}$ must be formed. (We follow in this paper the notation used by Eichhorn and Clary (1974a). There, however, our α is called Δ and our ξ is called v.) Note that this requires knowledge of approximate values α_0 for α. Approximate values for the x_0 are, of course, available – they are the observations themselves. Approximate values α_0 for α may be obtained in the classical way by regarding the right-hand sides of the condition equations (1) as normally distributed errors. The contemporary algorithms also take into account that the covariance matrix σ of the (x_v, y_v) or of the (ρ_v, P_v) is not necessarily diagonal.

(b) Especially when the object under consideration is very narrow, we have the case that the errors of the observations are not negligibly small compared with (at least some of) the adjustment parameters. This was pointed out earlier by Eichhorn and Clary (1974b). It is thus to be expected that the solution obtained from a linearized set of equations does not yield those values for α and ξ which actually minimize the quadratic form $\xi^T \sigma^{-1} \xi$. This requires either that second-order terms in the observation errors ξ be carried in the equations or, as Jefferys (1980) has pointed out, that iterations be performed using in the evaluation of the matrices x and $\mathbf{A}$ not only improved approximations for α as they become available, but also improved values for the observed quantities (i.e., $x = x_0 + \xi$, where the ξ are the negative adjustment residuals) as these become available. It is actually only then necessary to resort to carrying second order terms in the ξ when an approach with first order terms alone will not converge, or if other methods for solving the normal equations do not work (cf. Jefferys, 1981).

We can expect this modified Kowalsky–Seeliger algorithm to yield better values for the adjustment parameters α than the traditional approach to adjustment. Some of the criticism levied against this method may thus have to be revised. Note, however, that it can never be regarded as the basis for a definitive orbit determination since it makes no use of the observation epochs.

2. Critique of the Power-Series Methods

As reported in Heintz's (1971) monograph, a class of orbit determination methods was proposed in which position angles (as well as distances) are represented as polynomials in time. (In principle, this could also be done for rectangular coordinates.) These polynomials are generated as broken-off power series and usually go no higher than to t^3. Furthermore, the area theorem imposes conditions on the coefficients of the position angle series (polynomial) and the distance series (polynomial). Since the coefficients in these series can, in principle, be expressed in terms of the orbital elements, the orbital elements themselves may be computed after a sufficiently large number of coefficients have been determined, provided that the resulting system is not mathematically singular. Experience with actual numbers shows, however, that the coefficients of t^3 are usually not too well determined, especially for short arcs which, after all, are characteristic for those systems which still need to have their orbits determined. The orbital elements themselves can therefore not be found with high accuracy. Under the circumstances, this is to be expected.

Another serious drawback of this approach is the following. One assumes that the relationship between the coefficients of the approximation polynomial and the orbital elements is that which holds for the corresponding coefficients in a series development. These relationships can be derived straightforwardly. The approximation polynomial method, on the other hand, models actually the relationship between coordinates (either x_v, y_v or ρ_v, P_v) and time t_v to be of the form

$$P(t) = P(a, t) = P_0 + P_1 t + P_2 t^2 + P_3 t^3 \tag{4}$$

and an analogous expression for $\rho(t)$, subjected to those restrictions between the coefficients of P and of ρ which enforce the area theorem (a stands for the orbital elements). Equation (4) is, however, patently false, meaning that there are no P_v ($v = 0, 1, 2, 3$) which make this equation true. Equation (4) may still be used as equation of condition for the estimation of the coefficients $P_0, \ldots, P_3$ of the approximation *polynomial* (not series!). This would lead to a system of normal equations whose matrix would consist of sums of powers of t, and whose right-hand sides could be approximated by expressions of the form

$$P_v = \frac{1}{t_1 - t_0} \int_{t_0}^{t_1} t^v P(a, t) \psi(t) \, dt \quad (v = 0, \ldots, 3), \tag{5}$$

where t_0 and t_1 are the epochs of beginning and end, respectively, of the series of observations, and where $\psi(t)$ is a distribution function representing a combination of the time-dependence of the observation density and observation precision. The most realistic theoretical relationship between the coefficients P_v and a would be that found by solving the system of normal equations alluded to above. The relationships found in this way will the more deviate from those which a straight series development would yield, the more the observed orbit segment deviates from a circle. The verdict for short arcs has so far been that the traditional orbital elements simply cannot be found with high accuracy.

3. Satisfaction of the Dynamical Conditions

Consider a right-handed astrocentric coordinate system K such that its $x - y$-plane is the orbital plane and that the positive x-axis points toward the periastron (of the secondary with respect to the primary). The positive y-axis of K is obtained by rotating the system by $90°$ on the z-axis in the direction of the orbital motion. Consider a second astrocentric, right-handed system k whose axes are parallel to those of the equator system Q. The transformation between K and k is given by

$$x^k = \mathbf{R}_3(-\Omega)\mathbf{R}_1(-i)\mathbf{R}_3(-\omega)x^K . \tag{6}$$

From the theory of the two-body problem we know that the coordinates of the secondary with respect to the primary, in K, are given by

$$x^K = \begin{pmatrix} X \\ Y \\ Z \end{pmatrix} = a \begin{pmatrix} \cos E - e \\ \sqrt{1 - e^2}\, \sin E \\ 0 \end{pmatrix} , \tag{7}$$

where the eccentric anomaly E is the solution of Kepler's equation

$$n(t - T) = E - e \sin E , \tag{8}$$

n being the mean motion and T the periastron epoch, the two orbital elements not involved in Equation (1). From Equations (6) and (7) we get

$$x^{k\cdot} = \begin{pmatrix} x \\ y \\ z \end{pmatrix} = \begin{pmatrix} A & F & K \\ B & G & L \\ C & H & M \end{pmatrix} \begin{pmatrix} \cos F - e \\ \sqrt{1 - e^2}\, \sin E \\ 0 \end{pmatrix} , \tag{9}$$

where A, B, F, G are known as the Thiele–Innes constants. They are uniquely related to α and allow one, when known, to compute a, i, Ω, and ω, from which the other elements of the matrix in Equation (9) can be calculated because the matrix in Equation (6) is a times an orthogonal matrix. For this reason, we also have

$$\begin{pmatrix} \cos E - e \\ \sqrt{1 - e^2}\, \sin E \\ 0 \end{pmatrix} = \frac{1}{a^2} \begin{pmatrix} A & B & C \\ F & G & H \\ K & L & M \end{pmatrix} \begin{pmatrix} x \\ y \\ z \end{pmatrix} , \tag{10}$$

whence

$$z = \frac{-1}{M} (Kx + Ly) . \tag{11}$$

As it happens, $M = \cos i$. Inserting Equation (11) in Equation (9), we see that x^K and

y^K can be expressed in terms of a, i, Ω, ω, and the observations x and y.

The area theorem is expressed by the equation

$$X\dot{Y} - \dot{Y}X = na^2 \sqrt{1 - e^2}, \tag{12}$$

the right-hand side being the area constant.

We also know from the theory of elliptical two-body motion that

$$\begin{pmatrix} \dot{X} \\ \dot{Y} \\ \dot{Z} \end{pmatrix} = \frac{an}{\sqrt{1 - e^2}} \begin{pmatrix} -Y/\sqrt{X^2 + Y^2} \\ X/\sqrt{X^2 + Y^2} + e \\ 0 \end{pmatrix}. \tag{13}$$

$\dot{X}$ and $\dot{Y}$ can be expressed in terms of X and Y (as well as the elements a, n, and e) through Equation (13) and in turn, as functions of x, y, and the orbital elements through Equations (10) and (11). Inserted in Equation (12), this equation then expresses the area theorem exclusively in terms of the observed relative coordinates and the elements of the orbit. If one carries out the calculations, one sees that this makes basically the same statement as Equation (1), namely, that all points on the projected orbit lie on an ellipse. The observation epochs are still not used, the area theorem has not helped. We need an equation which is a relationship between x, y, and the observing time t.

This equation is established as follows.

From Equation (7) we see that

$$\cos E = X/a + e, \qquad \text{thus} \quad E = \text{arc } \cos(X/a + e) \tag{14}$$
$$\sin E = Y/(a\sqrt{1 - e^2}), \quad \text{thus} \quad E = \text{arc } \sin[Y/(a\sqrt{1 - e^2})].$$

Using the expressions for $\sin E$ and E from Equation (14) in Equation (8), we see that

$$\frac{X}{a} + e = \cos\left[n(t - T) + \frac{eY}{a\sqrt{1 - e^2}}\right], \tag{15}$$

which yields the required equation (after X and Y are expressed in terms of x and y) that must be used, in addition to Equation (1), as equation of condition for adjusting the observations x_ν and y_ν.

Note that the equations

$$H = \sin[n(t - T) + eH], \tag{16}$$
$$\Xi = \cos[n(t - T) + e\sqrt{1 - \Xi^2}],$$

with

$$\Xi = \frac{X}{a} + e, \qquad H = \frac{Y}{a\sqrt{1 - e^2}} \tag{17}$$

are transcendental equations in X and Y, and can be used to calculate X and Y directly as functions of time in the same way as Kepler's equation (8) is used to calculate E as a function of t. Likewise, we find that

$$H = \sin\left[n(t - T) + e\sqrt{1 - \Xi^2}\right]. \tag{18}$$

Equations (16) or (18) may be used in the same way as Equation (15) for enforcing the dynamical conditions to which the observed relative coordinates and the elements of the orbit are subjected.

4. Suggestions for Practical Applications

We see that the observed relative positions of the companion with respect to the primary (without restricting generality these may be regarded as the rectangular coordinates x and y) must simultaneously satisfy Equation (1) and one of the Equations (15), (16), or (18). Again without restricting generality, we may choose Equation (15) for this purpose.

In Section 1 above we have suggested how to achieve the best adjustment of the x, y when they are subjected to Equation (1) alone. This procedure yielded approximation values for a, e, i, Ω and ω. With these, the adjustment must be repeated, this time including Equation (15) in the set of condition equations. To do this, approximation values for n and T must also have been found.

The same remarks apply – *mutatis mutandis* – which were made above in Section 1 in connection with using Equation (1) as the only type of condition equations. It may be necessary to include second order terms in the parameter corrections as well as in the observing errors to achieve convergence.

The implementation of these suggestions is currently the subject of a dissertation (Yu-lin Xu) at the department of Astronomy at the University of Florida.

References

Eichhorn, H. and Clary, W. G.: 1974a, *Monthly Notices Roy. Astron. Soc.* **166**, 433.
Eichhorn, H. and Clary, W. G.: 1974b, *Monthly Notices Roy. Astron. Soc.* **166**, 425.
Heintz, W. D.: 1971, *Doppelsterne*, Wilhelm Goldmann Verlag, München.
Jefferys, W. H.: 1980, *Astron. J.* **85**, 177.
Jefferys, W. H.: 1981, *Astron. J.* **86**, 149.

DISTRIBUTION OF ASTROMETRIC BINARIES IN VARIOUS GALACTIC LATITUDES*

A. N. GOYAL

Department of Mathematics, Univ. Raj., Jaipur, India

(Received 31 July, 1984)

Abstract. The conclusions of the present paper broadly are: (a) The galactic concentration of doubles by comparing the distributions in galactic latitudes $0° < \beta \leq 20°$ and $\beta > 40°$ is nearly twice as large as the galactic concentration of stars in general. (b) The astrographic catalogues are not complete in the fainter magnitudes. (c) The large value of the ratio $T:O_k$ (observed to optical number of pairs) from Kreiken's formula shows that almost all stars in the group $0° < d \leq 5''$ and quite a few in the other two groups, viz., $5'' < d \leq 10''$ and $10'' < d \leq 15''$ might be shown true binaries. Consequently, Aitken's working definition of a true binary should be extended if it were to include all true binaries. (d) The doubles are probably stars of Population I. (e) The logarithm to the base 10 of the cumulative counts can be represented by an empirical relation $A + B(\Delta m - 1.5) + C(\Delta m - 1.5)^2$.

1. Introduction

Innes (1907) discussed the utility of astrographic catalogues in picking binaries with wide separations and concluded that fifty years, hence, when the sky is repeated and the pairs show changes in the positions it would be possible to compare the binaries of the Milky Way and some yellow stars which would be an important factor in the solution of many sidereal problems.

Plummer (1908) found that the reason for the reluctance of the double star astronomers to compare their micrometrical observations of separation and position angle with the ones obtainable from astrographic catalogues, perhaps, lies in the inacurate measurements and reductions. Consequently, by systematic computation and comparison of separation and position angle from measures given in astrographic catalogues from formulae obtained by him, Plummer (1908) was led to conclude: (a) Differential measures of pairs of stars can be obtained from published astrographic catalogues with very little trouble. (b) Such measures have an accuracy of the same class as the results of the very best visual observers. (c) The photographic magnitudes are at any rate not inferior to the estimates of the skilled double star astronomer. (d) If the foregoing conclusions are justified, some guidance may be obtained as regards the policy which the observer at the telescope should follow.

In general, the photographic method does not allow of the accurate measurements of very close companions. The limit of separation required depends on the magnitude of the components. Also, the faintest star lies outside the great mass of easily accessible data which the astrographic catalogues will contain. There is thus ample field in which

* Communication presented at the International Conference on 'Astrometric Binaries', held on 13–15 June, 1984, at the Remeis-Sternwarte Bamberg, Germany, to commemorate the 200th anniversary of the birth of Friedrich Wilhelm Bessel (1784–1846).

the work of visual observer will be complementary to the photographic results. However, the work of de Kort (1958) and Jonckheere (1954) has given an insight into an extraordinary number of mistakes made in the measurement and reduction of plates. But, nevertheless, care ought to be exercised not to trench on a ground where visual observations will be purely redundant. Thus, in the course of comparing *Potsdam Photographische Himmelskarte* with *Oxford Astrographic Catalogues* in declination zones $+32°$ and $+33°$ and with *Paris Astrographic Catalogues* in the declination zones $+34°$ and $+35°$ for listing stars with large proper motions (Goyal, 1958a, b, c, 1962a, b, 1964a, b, c, d, 1965a, b, 1967; Goyal and Sharma, 1961; Goyal *et al.*, 1961; Goyal and Mithal, 1964; Goyal and Sharma, 1964; Goyal and Shringi, 1965; Goyal and Chaturvedi, 1967; Khandelwal and Goyal, 1970a, b, 1971) the author has given 93 new doubles with common annual relative proper motions (Goyal, 1958a, b, c, 1962a, b, 1964a, 1966a, b; Goyal and Khandelwal, 1968a, b, c). The distributions of small residuals is discussed by Goyal (1960), Goyal and Sharma (1961), Goyal *et al.* (1961), Goyal and Khandelwal (1969). Also the B.D. star 32.2896 (72 Herculis) having visual magnitude 5.4 and photographic magnitude 6.0 occurring in two overlapping regions with centre R.A. and declination $17^{h}15^{m} +32°$ star No. 168 and $17^{h}19^{m} +33°$ star No. 60 having epoch difference 34.8 and 42.6 yr, respectively, gave a mean annual relative proper motion from two reductions as $U_{x_0} = 0\rlap{.}''153 \pm 0\rlap{.}''008$ (accepted $U_{x_a} = +0\rlap{.}''126$) and $U_{y_0} = -1\rlap{.}''025 \pm 0\rlap{.}''008$ (accepted $U_{y_a} = -1\rlap{.}''047$). The errors are probable errors obtained for a single proper motion from differences of overlapping regions for this series of plates. Thus, tentatively, there is enough evidence on the utility of astrographic catalogues.

Aitken (1911) for the first time laid down certain restrictions in the definition of a visual binary. The views of active double star astronomers are also printed there. Groot (1927) conducted a survey of Greenwich astrographic catalogues in the declination zones lying between $+64°$ to $+90°$. Using W. Struve's well-known formula

$$D = \frac{N(N-1)\pi d^2}{2a^2} ,\tag{1}$$

where D is the number of optical doubles of separation less than d'' on an area a squares seconds having N stars, for the computation of the number of doubles which are optical pairs, Groot (1927) concluded that Aitken's (1911) 'working definition' of a binary should be extended if it were to include all true binaries. Curtiss (1929) and Couteau (1960) have also written on double star numbers and separation limit. Father Stein (referred to in Goyal, 1962a, b) showed that a difference of 20% might originate in the number of optical doubles obtained from Struve's formula if there was a star cluster in the region under consideration. Kreiken (1928) in his several papers, analysed the double star catalogues complied by Scheiner (*Publicationen des Astroph. Obs. zu Potsdam*, No. 59), Father Stein (*Catalogo Astrografico Sezione Vaticana Appendix III*, V, 1958), and Groot (1927), taking into consideration the difference of magnitudes between the components as well as different magnitude limits. He used the formula

$$D_0(m) = (d^2/r^2)A(m)\left[\frac{A(m)-1}{2} + N(m_1) - N(m)\right],\tag{2}$$

where $D_0(m)$ is the number of optical pairs between the magnitudes $m - \frac{1}{2}$ to $m + \frac{1}{2}$, $A(m)$ the number of stars between the limits of apparent magnitudes $m - \frac{1}{2}$ and $m + \frac{1}{2}$, $N(m)$ the number of all stars brighter than a given limit m and $N(m_1)$ the number of stars down to the limiting magnitude of the catalogue. Similar formulae are given by Curtiss (1929) also. Kreiken's (1928) main conclusions briefly are: (a) The percentages of physical pairs decreases with increasing magnitude m and increasing magnitude difference Δm between the components. (b) The percentages of physical pairs in the higher latitudes were generally larger than in lower ones. (c) A large number of physical doubles have an angular separation greater than $5''$ (a conclusion previously arrived at by Father Stein and Dr Groot).

Incidentally, it may be mentioned here that Öpik (*Publications de l'Observatoire Astron. de Tartu*, Tome XXV, No. 6) has also discussed the m, Δm distribution of doubles according to the spectra. But the spectral sub-divisions were found to be impossible by Kreiken (1928) in his case. Further, Kreiken (1928) found that the galactic concentration of doubles is only slightly larger than the stars in general. But he concluded that 'it would be dangerous to conclude that the concentration of doubles towards the Milky Way is larger than the stars in general'. He cited this conclusion to be in clear contradiction to that of Aitken (1911). Three things stand out in Kreiken's (1928) work: (a) non-uniformity of the material, (b) the material was not a good sample of the sky, (c) defective system of weighting (which he admits). Consequently, a systematic search of astrographic catalogues for doubles under an angular separation less than $15''$ and their analysis was undertaken by the author (Goyal, 1962a, b; 1964a, b; 1965a, b, 1967; Goyal and Shringi, 1965; Goyal and Verma, 1966) on the declination zones given in Table I.

TABLE I

Declination zones	Name of observatory
$-17°$ to $-23°$	Nizamiah Observatory, Hyderabad
$+36°$ to $+39°$	Nizamiah Observatory, Hyderabad
$+25°$ to $+33°$	Oxford
$+34°$ to $+35°$	Paris
$-38°$ to $-40°$	Perth
$-71°$ to $-81°$	Melbourne
$-41°$ to $-51°$	Cape

The general conclusion arrived at in the papers referred to can be summarized as follows:

(a) The galactic concentration of doubles obtained by comparing the number of doubles in $0° < \beta \le 20°$ and $\beta > 40°$, β being the galactic latitude is greater than the galactic concentration of stars in general.

(b) Aitken's (1911) working definition of a binary should be extended if it were to include all true binaries.

(c) The astrographic catalogues are not complete in the fainter magnitudes.

TABLE II

d (the separation)	$T:O_k$ (observed to optical)
$0''-5''$	8.2097
$5''-10''$	2.3427
$10''-15''$	0.8570

In Table II the average ratios of the observed number of doubles to the optical number of doubles calculated from Kreiken's (1928) formula are given for various separation.

The ratio $T:O_k$ is quite large which is an indication that a good number of doubles in $0'' < d \le 10''$ will turn out true binaries. But in passing from $0'' < d \le 5''$ to $5'' < d \le 10''$ the fall in the ratio is sharp. Consequently, Aitken's (1911) working definition of a binary should be extended if it were to include all true binaries (a conclusion arrived at earlier also by the author in the papers referred to).

In the beginning it was considered necessary to reduce the magnitudes to the International Photographic Scale for the purpose of homogeneity. This required a long cycle of interpolations from star counts to Chapman–Melotte scale (1914) with the help of the table given by Turner (1915) then to International Photographic Magnitudes with the help of the table given by Seares, *et al.* (1925). This was tried for one zone of each observatory. The magnitude scales were found to be identical (Goyal, 1965a, b; Goyal and Shringi, 1965; Goyal and Sharma, 1964; Goyal and Khandelwal, 1969). Consequently, this long cycle of interpolations was done away with.

While counting the doubles many zeros occurred in the fainter magnitudes than 11.0 mag. over the running Δm, perhaps, because of the incompleteness of the astrographic catalogues. Similar situation occurred for $m = 12.0$ or 13.0 and $\Delta m = 1.2$ and 0.3, respectively. In order to avoid the influence of irregularities in the limiting magnitude, I think, while comparing the number of stars in various galactic latitudes, the best thing is to neglect such stars which are fainter than $13\overset{m}{.}0$ for $m + \Delta m$.

2. Galactic Concentration

The detailed observed distributions according to d, m, Δm, β are given for various catalogues in the papers referred to above (Goyal, 1962a, b, 1964a, b, c, 1965a, b, 1966a, 1967; Goyal and Shringi, 1965; Goyal and Verma, 1966). The approximate latitude was found with the help of Olsson's (1932) tables. In Table III I have given the ratio of the

TABLE III

d (separation)	$0° < \beta \le 20°/\beta > 40°$	$20° < \beta \le 40°/\beta > 40°$
$0''-5''$	4.9548	2.4380
$5''-10''$	6.6960	2.9200
$10''-15''$	10.8748	3.4416
$0''-15''$	7.1284	2.8781

TABLE IV

Zones	Galactic concentration of doubles	Galactic concentration of stars in general
Oxford	6.113	3.226
Hyderabad (+)	10.274	4.370
Hyderabad (−)	4.912	2.270 ⎱
		3.290 ⎰
Perth	8.000	3.870
Melbourne	5.270	3.049
Paris	11.947	4.264
Cape	5.436	3.246
Seares and Van Rhijn	−	3.9
Vatican	8.70	−

number of doubles for each value of d (separation) in $0° < \beta \le 20°$ to $\beta > 40°$ and $20° < \beta \le 40°$ to $\beta > 40°$.

In Table IV I reproduce some of the results given in Goyal (1964c).

It is evident from Table III and Table IV that the galactic concentration of doubles is at least twice as large as the galactic concentration of stars in general. The average from Table IV for doubles and stars in general works out to be 7.5815 and 3.4983, respectively, which give a ratio doubles to stars in general as 2.1672 nearly. Also from Table III above if we take the ratio of galactic concentration of doubles to galactic concentration of stars in general in the separation group 0″ to 15″ we arrive at 2.4767 and, thus, the agreement appears to be good. But now the question is, if the galactic concentration of doubles is larger, then doubles must be stars of Population I. Further, researches on this aspect of the problem by astronomers who have the means for it would enable us to resolve this problem. The large difference of value in the galactic concentration is sufficient proof, I think, to show that doubles have a greater galactic concentration than the stars in general, and should it be true then a possible explanation could be that doubles are stars of Population I with a tendency to stay near the Galaxy (a conclusion arrived at earlier in Goyal, 1964a, b, c; Goyal and Mithal, 1964).

3. On the Cumulative Counts of Doubles

An empirical representation of the observed distribution of doubles under an angular separation less than 15″ in various galactic latitudes have been discussed by Goyal and Khandelwal (1968b). The cumulative counts of the observed distribution of doubles in different d in each astrographic zone were recorded. The average of these cumulative counts was obtained in each three astrographic zones in the case of O.A.C. and H.A.C. only. Representing the logarithms to the base 10 of the cumulative counts in the interval of the magnitude difference $0″.0 < m \le 2.7$ for average magnitude of the primary to be $11″.0$ by an empirical relation

$$\log_{10} N_m = A + B(\Delta m - 1.5) + C(\Delta m - 1.5)^2 \tag{3}$$

and solving by the method of least squares we obtain

	A	B	C
$0'' < d \leq 5''$	1.2790 ± 0.0461	0.1994 ± 0.0214	$-(0.1644 \pm 0.0219)$
$5'' < d \leq 10''$	2.0655 ± 0.0416	0.2305 ± 0.0174	$-(0.1310 \pm 0.0121)$
$10'' < d \leq 15''$	2.0760 ± 0.0395	0.2317 ± 0.0116	$-(0.1192 \pm 0.0086)$

Acknowledgements

My thanks are due to several co-workers who have worked hard during the initial stages of several papers included in the references. I am specially thankful to Dr R. S. Khandelwal and Mr S. S. Mittal, my collegues in the Department of Mathematics, who have substantially contributed in the preparation of the paper.

References

Aitken, R. G.: 1911, *Astron. Nachr.* **188**, 281.
Chapman, S. and Melotte, P. J.: 1914, *Mem. Roy. Astron. Soc.* **60**, 000.
Couteau, P.: 1960, *J. Obs. Marseille* **43**, 41.
Curtiss, R. H.: 1929, *J. Roy. Astron. Soc. Canada* **23**, 422.
De Kort, J.: 1958, *Catalogo Astrografico Sezione Vaticana*, Appendix V.
Goyal, A. N.: 1958a, *J. Obs. Marseille* **41**, 21.
Goyal, A. N.: 1958b, *J. Obs. Marseille* **41**, 121.
Goyal, A. N.: 1958c, *Astrophys. J.* **63**(4), 147.
Goyal, A. N.: 1960, *Proc. Raj. Acad. Sci. Pilani India* **7**, 1.
Goyal, A. N.: 1962a, *A.N. Band* **286**, Heft 5, p. 196.
Goyal, A. N.: 1962b, *Monthly Notices Roy. Astron. Soc.* **123**, 413.
Goyal, A. N.: 1964a, *Monthly Notices Roy. Astron. Soc.* **127** (3), 215.
Goyal, A. N.: 1964b, *Monthly Notices Roy. Astron. Soc.* **127** (3), 341.
Goyal, A. N.: 1964c, *Monthly Notices Roy. Astron. Soc.* **128**, 493.
Goyal, A. N.: 1964d, *J. Obs. Marseille* **47**, 221.
Goyal, A. N.: 1965a, *J. Obs. Marseille* **48**, 158.
Goyal, A. N.: 1965b, *Proc. Nat. Inst. Sci. India* **31**(A), 33.
Goyal, A. N.: 1966a, *Monthly Notices Roy. Astron. Soc.* **133** (1), 15.
Goyal, A. N.: 1966b, *Proc. Nat. Acad. Sci. Allahabad India* **36**(A), 510.
Goyal, A. N.: 1967, *Proc. Nat. Inst. Sci. India* **33**(A), 80.
Goyal, A. N. and Chaturvedi, K. K.: 1967, *J. Obs. Marseille* **50**, 101.
Goyal, A. N. and Khandelwal, R. S.: 1968a, *Astrophys. J.* **68**, 366.
Goyal, A. N. and Khandelwal, R. S.: 1968b, *Proc. Nat. Inst. India* **34**(A), 36.
Goyal, A. N. and Khandelwal, R. S.: 1969, *Proc. Nat. Inst. India* **35**(A), 434.
Goyal, A. N. and Mithal, S. S.: 1964, *Proc. Nat. Acad. Sci. Allahabad India* **34**(A), 358.
Goyal, A. N. and Sharma, B. L.: 1961, *Vikram India* **5**, 46.
Goyal, A. N. and Sharma, B. L.: 1964, *Publ. Astron. Soc. Japan* **16**, 67.
Goyal, A. N. and Shringi, P. C.: 1965, *Publ. Astron. Soc. Japan* **17**, 66.
Goyal, A. N. and Verma, S. K.: 1966, *Proc. Nat. Inst. Sci. India* **31**(A), 146.
Goyal, A. N., Sharma, B. L., and Srivastava, V. K.: 1961, *Vikram India* **5**, 133.
Groot, H.: 1927, *Monthly Notices Roy. Astron. Soc.* **88**, 51.
Innes, R. T. A.: 1907, *Observatory* **390**, 452.

Jonckheere, R.: 1954, *J. Obs. Marseille* **37**, 31.

Khandelwal, R. S. and Goyal, A. N.: 1970a, *Monthly Notices Roy. Astron. Soc.* **150**, 235.

Khandelwal, R. S. and Goyal, A. N.: 1970b, *Indian J. Pure Appl. Maths.* **1**(2), 192.

Khandelwal, R. S. and Goyal, A. N.: 1971a, *Indian J. Pure Appl. Maths.* **1**(3), 284.

Khandelwal, R. S. and Goyal, A. N.: 1971b, *Indian J. Pure Appl. Maths.* **2**(2), 155.

Kreiken, E. A.: 1928, *Bull. Astron. Inst. Netherlands* **4**, 405, 212, 219, 239.

Olsson, John.: 1932, *Lund Univ. Annals* **3**, 1932.

Plummer, H. C.: 1908, *Monthly Notices Roy. Astron. Soc.* **69**, 100.

Seares, F. H., Van Rhijn, P. J., Joyner, M. C., and Richmond, M. L.: 1925, *Astrophys. J.* **62**, 320.

Turner, H. H.: 1915, *Monthly Notices Roy. Astron. Soc.* **75**, Table II.

ASTROMETRIC BINARIES AND THEIR ROLE IN DOUBLE STAR ASTRONOMY*

T. HERCZEG

University of Oklahoma, Norman, Oklahoma, U.S.A.

(Received 15 October, 1984)

Discussing the relationship between the astrometric binaries and the family of double stars is somewhat like dealing with the 'philosophy' of this type of binaries. Available evidence points toward the assumption that we encounter here *primarily* a specific technique of observation rather than a specific type of objects. This technique addresses itself to moderately wide pairs among nearby stars, with a in the 'grey area' around 10 AU, objects whose detection by other methods of observation presents difficult problems. In this context astrometric binaries may be considered an extention of the class of visual binaries, as illustrated by systems where the perturbing unseen component has been rendered visible by more powerful instruments, as in the case of Sirius B, predicted by Bessel, Procyon B, and Ross 614.

The limitations of the techniques are well known: pairs with $\Delta m \approx 0$ are basically undetectable; cases with larger Δm values become increasingly difficult, as the photocentric orbits grow smaller and for $\Delta m > 5$, will be reduced to the absolute orbits of the more massive – in this case much more massive – components; finally, if $r \gtrsim 20$ pc, for all systems a and, hence, α (the semi-major axis of the photocentric orbit) becomes very small. In discussing the detectability of an astrometric system, one should bear in mind the important figure giving the available accuracy, as it emerges from numerous solutions worked out by P. van de Kamp and his collaborators: the probable error of α is of the order of $\pm 0\overset{''}{.}004$ or $\pm 0\overset{''}{.}005$, the accuracy of a good trigonometric parallax. Based on this figure, it is very easy to construct charts of detectability, depending on the orbital elements of a system and its distance from the Sun. It should be added, in appreciation of the great amount of work required, that for most of the detectable systems the periods fall in the range of 10 years to 50 years, that is, a sustained, patient effort is necessary to study them, preferably with the same instrument. This type of research is certainly little suited to 'guest investigations' – a point possibly important for accommodating future studies on satellites. On the other hand, there are indications that ground-based methods are just before a break-through, with improvements raising the accuracy of the best individual measurements (of the relative position) from about $0\overset{''}{.}01$ to $0\overset{''}{.}001$.

Attempts at determining the space density of astrometric binaries should be based on estimating, from known distributions of stellar data (such as the luminosity function,

* Communication presented at the International Conference on 'Astrometric Binaries', held on 13–15 June, 1984, at the Remeis-Sternwarte Bamberg, Germany, to commemorate the 200th anniversary of the birth of Friedrich Wilhelm Bessel (1784–1846).

the rate of duplicity, the distribution of mass-ratios) the expected number of binaries in the characteristic range of orbital elements and parallax. This approach emphasizes the obvious assumption that in most cases of unseen companions we deal with members of 'ordinary' binary systems. The opposite approach, over counting the unseen components in their distance dependence and establishing an $N(r)$ distribution, is of little avail; thus the occasionally quoted results of a surprisingly high number (and mass) density in space of astrometric binaries is to be considered with caution. The weakness of this procedure is that no $N(r) \sim r^3$ plateau can be established in the counts, the apparent numbers fall off sharply from the very beginning, already after $r \lesssim 1$ pc.

On the other hand, the probably yet incomplete data for the region within 5 pc from the Sun suggest that about 20% of the single stars and additionally one or two double and multiple systems may have unseen companions. This is not far from the number what we may have to expect from the properties of the stellar sample in the solar neighborhood, representing predominantly the lower Main Sequence.

The earlier statement that most astrometric binaries are stellar pairs in the range of separation of 5 to 15 AU, should not obscure the fact that a few systems may still represent new types of objects harboring, in particular, possible black or gray dwarf components (masses: 0.01 $M_\odot$ to 0.06 $M_\odot$) or massive, Jupiter-size planets. Here the role of astrometric methods is probably crucial. Concerning black dwarfs, estimates show a fair chance of detectability within 5, even 10 pc distance, at least for not too close components, $a \geq 5$ AU. Among about 30 astrometric systems discussed in the literature, only 4 or 5 cases may show the presence of possible substellar masses – yet none of them is undisputed. There may be found one or two cases of truly substellar masses, with perhaps BD 43°4305 and Stein 2051 as the best candidates. However serious the observational selection effects are in this small-number statistic, the preliminary indications are that these very low mass objects are not common components. Nevertheless, the systematic study of nearby M-stars may bring results of considerable interest.

As to possible planets around nearby stars, chances of detection ($\alpha \sim 0\rlap{.}''005$ to $0\rlap{.}''010$) are restricted to our immediate neighborhood, $r \lesssim 3$ pc, and low mass late M-type primaries.

These limitations and the still unsettled controversy about the planet or planets orbiting Barnard's star, underline the necessity of extending the search for extra-solar system planets to quite different methods, such as monitoring very small periodic variations of the radial velocity or direct viewing from above the terrestrial atmosphere.

Selected References

Abt, H. and Levy, S.: 1976, *Astrophys. J. Suppl.* **30**, 241.
Halbwachs, J. L.: 1983, *Astron. Astrophys.* **128**, 399.
Heintz, W. D.: 1978, *Astrophys. J.* **220**, 931.
Herczeg, T.: 1984, *Astrophys. Space Sci.* **99**, 29.
van de Kamp, P.: 1975, *Ann. Rev. Astron. Astrophys.* **13**, 295.
van de Kamp, P.: 1981, *Stellar Paths*, D. Reidel Publ. Co., Dordrecht, Holland.

WORK ON ASTROMETRIC BINARIES IN CHINA*

L. YAN, L. WAN

Shanghai Observatory, Academia Sinica, China

and

S. GONG

Purple Mountain Observatory, Academia Sinica, China

(Received 13 June, 1984)

Abstract. The investigations of astrometric binary systems, currently carried out in the People's Republic of China, are described and discussed.

1. Introduction

In the early 1900's the Zô-Se Observatory (the Sheshan Station of Shanghai Observatory since 1962) engaged in astrometry of visual binaries with its 40/714 cm twin refractor. Apart from the discovery of new ones, 1122 pairs of J. Herschel binaries were remeasured. In 1960, the Sheshan Station resumed its work on visual binaries. New orbital elements of visual binaries were computed and the method of orbital determination was investigated. 108 pairs of visual binaries have been discovered with the twin refractor.

At present, while continuing the measurement of visual binaries, Yan *et al.* have collected the elliptic orbital elements of more than 700 pairs of visual binary stars. By means of these elements, the positions are predicted from 1984 to 2003 apparent elliptic orbits are plotted in a rectangular coordinate system with the origin at the primary star, and relative positions of the secondary star are denoted at different time by the calculated orbital ephemerides.

A speckle interferometer attached to the 1.56 m astrometric reflector to measure binaries is being planned at Shanghai Observatory. At the Yunnan Observatory, the same is being made for the measurement of stellar angular diameters as well as of the separation of double stars.

2. Determination of Trigonometric Parallaxes of Selected Double Stars with Appreciable Orbital Motion

The primary research instrument at the Sheshan Station of the Shanghai Observatory will be a 1.56 m astrometric reflector ($f = 15\,600$ mm, scale value $13''.22$/mm), which was designed for extreme mechanical and thermal stability.

---

* Communication presented at the International Conference on 'Astrometric Binaries', held on 13–15 June, 1984, at the Remeis-Sternwarte Bamberg, Germany, to commemorate the 200th anniversary of the birth of Friedrich Wilhelm Bessel (1784–1846).

Astrophysics and Space Science **110** (1985) 135–136. 0004–640X/85.15
© 1985 *by D. Reidel Publishing Company.*

The telescope was designed to allow photographic measurements of binary stars. The optical calculation showed that the maximum stellar image within a field of $\pm$ 10' is less than 1". The maximum monochromatic light patch is 0".073 in the wavelength range from 4047 to 6563 Å. The primary and secondary mirrors are made of ultra-low-expansion Cer-Vit glass and have been ground by the Shanghai Institute of Optics and Fine Mechanics; the finished mirror is of excellent quality for photographic observations of binary stars.

Future programs at the Sheshan Station with the 1.56 m astrometric reflector will be extended to include the determination of trigonometric parallaxes of selected double stars with appreciable orbital motion.

3. Photoelectric Observations of Stellar Occultations

The occultation technique is presently a useful tool for double star research. The 40 cm, $f/17$ refractor at the Sheshan Station is used for this observation. There are pulse amplifier, discriminator and pulse counter. The integration time is 1 msec. The data-out is printed with a high-speed typewriter. The clock time is established with reference to the coordinate time signals by means of a radio set.

The equipment will be put into routine program observations, including the detection of double and multiple stars.

Acknowledgement

We wish to thank the United States Naval Observatory for the occultation predictions.

CLOSE TRIPLE STAR SYSTEMS*

FRANK BRADSHAW WOOD

Department of Astronomy, University of Florida, Gainesville, U.S.A.

(Received 13 June, 1984)

Abstract. Triple star systems, especially those in which one star has very small mass may be more common than has been generally considered. Here is summarized some of the recent evidence supporting this possibility.

This paper will not discuss astrometric binaries in the conventional sense, but rather will assume that we are chiefly interested in very close binary or multiple stars no matter how they are studied, and will actually concentrate more on triple-star systems than on binaries. A good deal of interesting work has been done lately along these lines and it will not be possible to cover all of it.

We will also try to avoid highly specialized topics such as details of methods of solutions of light curves of eclipsing variables. Nearly forty-five years ago it was written (cf. Dugan, 1939) that... 'It has often been stated that the eclipsing variable is only accidentically a variable star and that the problem it presents is explained by a little simple geometry'; the author immediately went on to point out the number and complexity of the papers being published even then, to explain this 'geometry'. The situation now is, of course, very much more complex, and has been discussed in a number of symposia and colloquia in recent years.

What we will try to do here is to see what kind of observations might tell us more about the physical nature of close systems. We will start with a very brief discussion of changes of the period of orbital revolution.

The fact that many close binary stars exhibit changes in their periods of revolution is now extremely well established, although the concept did not come easily. Perhaps naturally, the idea was accepted only slowly and some of the early suggestions now seem a bit strange to us. For example, a mechanism considered seriously in at least one case was a sudden encounter with an interstellar cloud.

The detection and study of changes in the evolution period is probably best obtained by observations of the precise times of light minima in eclipsing systems. The conventional method has long been a consideration of a plot of the (O–C)'s – i.e., the difference between the times of minima actually observed and those computed, usually

* Communication presented at the International Conference on 'Astrometric Binaries', held on 13–15 June, 1984, at the Remeis-Sternwarte Bamberg, Germany, to commemorate the 200th anniversary of the birth of Friedrich Wilhelm Bessel (1784–1846).

Astrophysics and Space Science **110** (1985) 137–141. 0004–640X/85.15
© 1985 *by D. Reidel Publishing Company.*

from a set of linear 'light elements' consisting of the initial epoch (one supposedly well determined time of minimum usually observed early in the study of the system) and the period of the system multiplied by an appropriate number of revolutions. In many cases, of course, the initial epoch can be the average of several observed ones made at about the same time, but this is a detail we need not worry over now. If the period is constant and the value is reliable, the (O–C)'s will scatter about the value of zero; if the period is in error they will depart in a linear manner; the direction and magnitude of such changes will given an indication of the corrections which must be applied to obtain a better value of the period.

However, in many cases the period does not remain constant and the nature and magnitude of its changes can tell us something about the system. It has long been customary to divide these systems first into those in which the period was constant and those in which it was not, although even here the more cautious observer used the term 'probably constant' or its equivalent. For example, in a study of AR Lacertae published in 1939, Dugan concluded that the observations 'give no evidence that the period is anything but linear'. Yet, even as these words were being written photoelectric observations were being made which would find the observed minima almost precisely an hour from the predicted value.

Further subdivision is usually made into groups in which the period changes are themselves periodic and those in which they seem to be sudden and unpredictable; the first group again is usually subdivided into those in which changes in the times of secondary minimum are the same as those in primary and those in which these are 180° out of phase.

It has long been known that apparent period changes can be caused by the presence of a third companion. The reasoning is simple. The eclipsing pair moves away from us on its motion around the common center of mass, so the period is apparently lengthened; later it is moving toward us, the period apparently decreases. The secondary is altered in the same manner; this helps to distinguish it from the case in which the periodic change of period is caused by the rotation of the line of apsides in a two-body system.

The present paper will discuss primarily three-body systems in which the third body cannot be studied by visual means. A few examples of recent work will be given. Nevertheless, it should be noted that period studies can 'destroy' existing stars as well as discover new ones. A paper on Algol by Frieboes-Condé *et al.* (1970) is of considerable interest here.

These authors carried out an investigation to clarify the structure of the system with special respect to a possible unseen companion. They rediscussed the extended series of astrometric observations using infrared spectroscopic elements for the triple system, Algol AB–C. Little change was found in the 1.862 yr triple star orbit but a combined solution for these and a hypothetical 32-yr periodicity showed that the latter does not correspond to orbital motion. A negative result was also found for the 'great inequality' in the observed times of minimum (period about 180 yr). The available meridian observations also exclude the possibility of a linear orbital motion, with the motion of the triple system remaining virtually rectilinear.

After excluding these two possible unseen companions and thus their light-time effects they were able to obtain a good representation of the photoelectrically determined epochs by two abrupt changes of the eclipsing period, one in 1944 and on in 1952. The size of the changes are $+3.5$ s and -2.1 s, respectively. On those sudden changes, there is superposed the 32-yr periodic variation and the 1.862-yr light-time effect. Incidentally, Algol is the system for which at least three observers reported minima in the blue coming from 16 to 24 min later than in the yellow. This so-called Tikhoff–Nordmann effect is no longer considered real, especially after later work by J. S. Hall and by R. Szafraniec, but there was a time when it attracted considerable attention.

A more recent discussion of period changes caused by a third companion has been given by Havnes (1980). This paper also gives excellent summaries of the recent work in the field.

Havnes notes that W UMa type binaries exhibit period changes of $\sim 10^{-6}$ days at rather short time intervals. The time intervals between these sudden period jumps may be of the order of five years. It is suggested that the changes could be caused by the periastron passage of a third body body moving in an orbit of high eccentricity. Restriction on the third body's mass make it unlikely that these are produced directly by dynamical action as secondary effects. Smaller period jumps could trigger mass transfer and ejection events which cause the observed change. It might be noted that more than thirty years ago it was pointed out (cf. Wood, 1950) that systems showing constant periods over relatively long intervals had both components well removed from the zero-velocity or Jacobian surface, while systems showing sudden changes had at least one component near this surface. It was also suggested that violent prominence activity might be responsible.

It was also noted by Havnes that J. A. Kreiner (in *IAU Colloq.* **42**, 343) found that changes take place in intervals of 5–25 years in between which the period is constant. Suggested causes of period changes include mass transfer within the system, mass ejection, perturbation of gas around the binary system, and magnetic braking due to stellar winds.

Havnes then finds it possible to produce changes by action of a third component in a highly eccentric orbit. Note that if the orbit is of low eccentricity this result will be the conventional light-time effect. Other authors (e.g. Whitmire and Motese, 1982) have suggested that alternate period changes could be the result of small fluctuations in stellar structure, such as a variation in stellar radius of about 3%.

Another extremely different system is Cyg X-1. This has recently been discussed by Manchanda (1983). He finds 'possible' evidence of a 300 day period in hard X-ray intensity from analysis of data running from 1964 through 1978. He suggests that this is due to the presence of a third body which will also explain the problems of the large mass assigned to the X-ray source. This source seems to be highly variable on different time-scales. This consists of (1) high-energy flaring activity and (2) rapid variations in the high-low state. The mass of the compact object is estimated at 8–15 times the mass of the Sun.

Another recent discovery is that EE Pegasi is a triple star. The discovery is due to

C. H. Lacy and D. M. Popper and I am greatly indebted to them for communicating the information in advance of publication, although this was to appear in the June issue of the *Astrophysical Journal* and thus is now available. Their discovery is the presence of a previously unknown faint companion. This has an orbital period of revolution of about 4 years, and has remained unsuspected until the present study despite the large amount of photometric and spectroscopic work done on this system since its discovery as an eclipsing pair in 1935.

Recent spectroscopic observations were obtained with the Reticon spectrometer of the McDonald Observatory 2.7 m telescope. The hope was that weak lines of the secondary could be observed in the 6400 Å region and these were indeed easily found. However, it was found that the velocity at the center of mass of the eclipsing system was modulated by the four year orbit about the triple star system center of mass. The same modulation was present in residuals of secondary as well as primary (and in the primary velocities found earlier by Popper) thus strenghtening the idea of a light-time effect.

The authors find that the minimum mass of the third body is 0.3 times the mass of the Sun as found from the mass function and the mass of the eclipsing pair; the maximum mass is more difficult to estimate. This emphasizes the importance of studies of this kind in detecting very low-mass objects. However, they feel it is 'conservative' to say that the third component would contribute less than 2% of the total light of the system.

Another system which now seems to have a low mass third companion is the very interesting system λ Tauri. Its discovery as a variable actually goes back to 1848, and its times of minima have been measured since 1855. A complete description of all the work done on it would be a lengthy process indeed; even the photoelectric observations begin in 1922. The first photometric 'solution' was published in 1950. Of course, the earlier observations come under some suspicion. This was one of the systems in which the so-called Tikhoff–Nordmann effect was found with the times of minimum in blue light arriving twenty minutes later than those in yellow. Much of the pertinent material has been recently summarized by Fekel and Tompkin (1982) who also added significant contributions of their own. These observers used low-noise Reticon spectra to re-measure the velocity curve of the primary and, for the first time, to obtain a velocity curve of the secondary. Masses of approximately 7.2 and 1.9 solar masses were determined.

A previously reported 33 day periodicity in the residuals was confirmed and, for the first time this was determined for both components so that orbital motion around a third body could be considered as definite. The mass of this third body was relatively small; the value given was 0.7 ± 0.2 solar masses.

The authors point out that this third component might well produce perturbations in the motion of the closer pair and suggest a precession period of the line of nodes of about seven to eight years – in line with previous suggestions – but point out the need for a new set of 'high quality' observations to settle the point. There is a good deal more to their discussion and analysis, but this summary should at least give some idea of the information available in such a study. Putting the 8 : 1 period ratio of this stellar system in the context of the solar system, they imagine the Sun as replaced by a binary and the Earth in place of the third stellar component. In this case, the Earth's orbit would be stable and thus, if dynamical stability alone is considered, 'the percentage of binaries that have planetary systems may be larger than is commonly supposed'.

Fairly recently, Dvorak (1979) raised an interesting question concerning the star HD 3765. "Is this", he asked, "an eclipsing binary star or is it an eclipsing planetary star?" This star was being used as a comparison star for observations of EG Andromedae. As is so often the case, the comparison star itself turned out to be variable. The plot of the observations showed a flat-bottomed minimum which could be caused by the total eclipse of a faint component or by transit of a small 'object' passing between us and the primary component. In either case, the duration of the entire eclipse $D_1 = 1\overset{d}{.}3$; the duration d of the flat bottom is $0\overset{d}{.}85$. As is well known, these can give at least an approximation to the ratio of the radii k, which in this case is less than 0.2. However, the parallax is also known for this system. Thus, the absolute magnitude can be computed and the radius determined from the Stefan–Boltzmann law. After some calculations it was found that $R_1 = 0.70$ solar radii and R_2 is less than 0.14 solar radii.

Dvorak first considered the possibility of a total eclipse. This means that a star of type K3 V eclipses a smaller, fainter star. The absolute magnitude of the secondary component is computed to be about + 10 so it can be a white dwarf or more probably a type-M subdwarf.

If we consider the case of a transit, it can be shown that the radius of the transiting object can be at the most only about 1.5 that of Jupiter and can be treated as that of a planet (or megaplanet).

The lack of more data, at the time of the paper did not permit a definite decision and, as is so often the case, the only firm conclusion was that more and better observations were needed.

In summary, perhaps the chief conclusion is that triple systems may be more common than has been assumed in the past and that the search for them by changes in motion is very close systems may be of considerable value in looking for stars of unusually small mass.

Thus we have studied 'astrometric binaries' by studying the changes in positions of the orbit of 'more conventional binaries'.

References

Dugan, R. S.: 1939, *Contr. Princeton Univ. Obs.*, No. 19.
Dvorak, T. Z.: 1979, *Acta Astron.* **29**, 151.
Fekel, F. C. and Tompkin, J.: 1982, *Astrophys. J.* **263**, 289.
Frieboes-Condé, H., Herczeg, T. J., and Hog, E.: 1970, *Astron. Astrophys.* **4**, 78.
Havnes, O.: 1980, *Astron. Astrophys.* **92**, 151.
Manchanda, R. K.: 1983, *Astrophys. Space Sci.* **91**, 455.
Whitmire, D. P. and Motese, S. J.: 1982, *Bull. Am. Astron. Soc.* **14**, 878.
Wood, F. B.: 1950, *Astrophys. J.* **112**, 196.

LUMINOSITY DISTRIBUTION FUNCTION FOR VISUAL BINARY STARS

KARL D. RAKOS

Institute für Astronomy, Wien, Austria

(Received 13 June, 1984)

Abstract. The 'cosmic scatter' of 147 of the best known visual binary stars on the Main Sequence is discussed and a new estimation of the luminosity distribution function for multiple star systems is presented. As long as the mass ratio q of a close binary is not smaller than 0.5, the distribution of close binary components is identical to the van Rhijn luminosity-function. For smaller mass ratios ($q < 0.5$) the number of close companions decreases rapidly. It appears that less than 13% of visual binaries in our sample are simple binary systems.

1. The Observational Material

Visual binaries were used by Murphy (1969) for a new Main-Sequence calibration of B stars. Stephenson (1960) studied the luminosity of G and K stars above the Main Sequence and Eggen (1963, 1965) used photometric data on binary stars to investigate the luminosities of all types of stars. However, the best known binary systems were not used for the investigations until now. The components are too close together and, therefore, not suitable for classical photoelectric photometry. In particular the complete lack of $B - V$ colors for components closer than 5 arc sec prevents the use of this group of stars – usually provided with reasonable orbits and distances – for the calibration. For the past few years ago it has been possible, by means of the area scanner technique, to measure the difference in brightness for binaries as close as 0.5 sec of arc with a high degree of accuracy. A total of 147 Main-Sequence double stars observations of this type published by Rakos *et al.* (1982), Hurly and Warner (1982) at Cape Town, and Lutz (1971) at Kitt Peak National Observatory, have been selected for analysis.

In general the standard deviation for typical $B - V$ value does not exceed 0.05 magnitudes. The same is true for the difference of the V magnitudes of the components. The accuracy is an order of magnitude higher than the cosmic scatter in the Main Sequence. The quality of the calibration, therefore, depends on the number of stars (good statistical sample) used in the analysis and not on the accuracy of the individual observations.

2. Calculation Procedure

Several binary systems are composed of components of the same colors. These stars provide direct information on the cosmic scatter. The modern theory of star formation

* Communication presented at the International Conference on 'Astrometric Binaries', held on 13–15 June, 1984, at the Remeis-Sternwarte Bamberg, Germany, to commemorate the 200th anniversary of the birth of Friedrich Wilhelm Bessel (1784–1846).

predicts that multiplicity should be the natural configuration in which stars are found. The observed binary frequency along the Main Sequence is one of the interesting parameters useful in obtaining the luminosity distribution function for multiple star systems. This function is very poorly known, and the results of several investigations – due to extreme celestion effects – are still controversial, but it is one of the most important touchstones for the theory of star formation.

In our sample, 46 binaries have components of the same $B - V$ colors, that is, the difference in the colors is less than or equal to $0\overset{m}{.}03$. From the distribution of ΔV as a function of $B - V$ for these stars it appears that the scatter is homogeneous for all values of $B - V$ between A0 and K7. The color versus brightness difference, Figure 1,

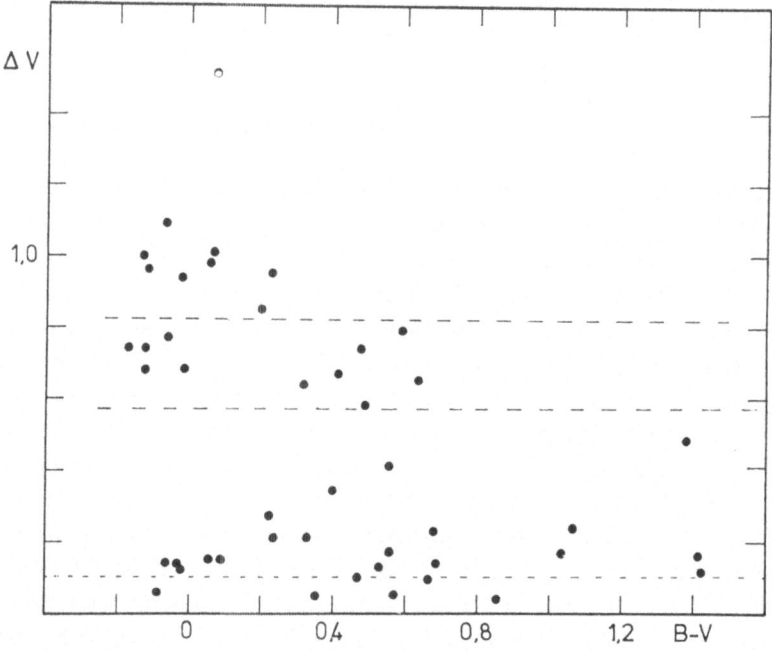

Fig. 1.   The color versus brightness difference for 46 binaries with components of the same $B - V$ colors.

and the histogram, Figure 2, show three different groups of stars. Stars with $\Delta V > 0\overset{m}{.}8$ are hot and evolved from the Main Sequence. The group around $\Delta V = 0\overset{m}{.}7$ are visual binaries and one of the binary components has a close unresolved companion of the type $AB–C$, all three components are of equal colors. In the third group the stars with $\Delta V < 0\overset{m}{.}1$ are simple binaries and the deviation in the brightness is introduced by the observing errors, or of intrinsic origin. For $\Delta V > 0\overset{m}{.}1$ with the peak about $0\overset{m}{.}15$ we have triple systems but the contribution of the small companion to the color of the unresolved system does not exceed $0\overset{m}{.}03$. This third group has the same structure of the histogram for hot and cool stars and is not influenced by the evolutionary effects.

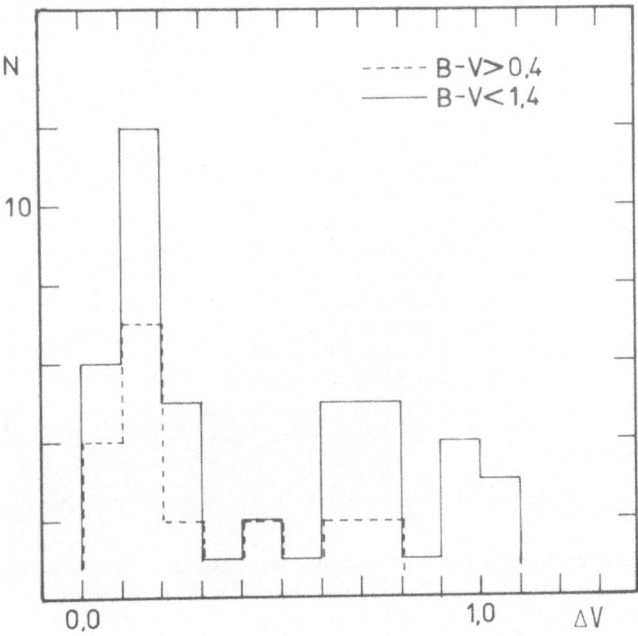

Fig. 2.   The histogram for the binaries in Figure 1.

A similar histogram as in Figure 2 can be constructed for all 147 stars used in this investigation placing all secondaries on the Main Sequence and considering the absolute values of the differences between the brightness of primaries and the Main Sequence. This method is not new. It was first proposed by Haffner and Heckmann in the 1930's and rediscussed in the 1970's by various authors, who applied it to the Main Sequence of open clusters. The histogram, Figure 3c, shows two features, one peak in $|\Delta M| = 0''.15$ and the broad accumulation around $|\Delta M| = 0''.7$. We would expect the position of the peak to be around $0''.0$, since components of a binary system have the same chemical composition. The result is not very sensitive to the color of stars. The histogram for hotter stars (secondaries), $B - V < 0''.3$ in Figure 3a, is similar to the histogram for late-type stars, $B - V > 0''.3$ in Figure 3b. For a few stars in the histogram the difference $|\Delta M|$ is larger than $0''.8$ and should be attributed to the evolutionary effects.

In general, the histograms in Figures 2 and 3 fit well together, considering the selection effect implemented in the first histogram. If we assume that one of the binary components has a close unresolved companion, and that the triple structure $(AB–C)$ is responsible for the ΔM deviation from the Main Sequence, we can estimate the luminosity – or mass – distribution of unseen companions B. It can be shown also that using the van Rhijn luminosity function for this distribution we are not able to reproduce the histogram in Figure 3c.

The van Rhijn function has its maximum for stars of 14th absolute magnitude, and in this case our histogram, regardless of the color of stars used, would have its maximum

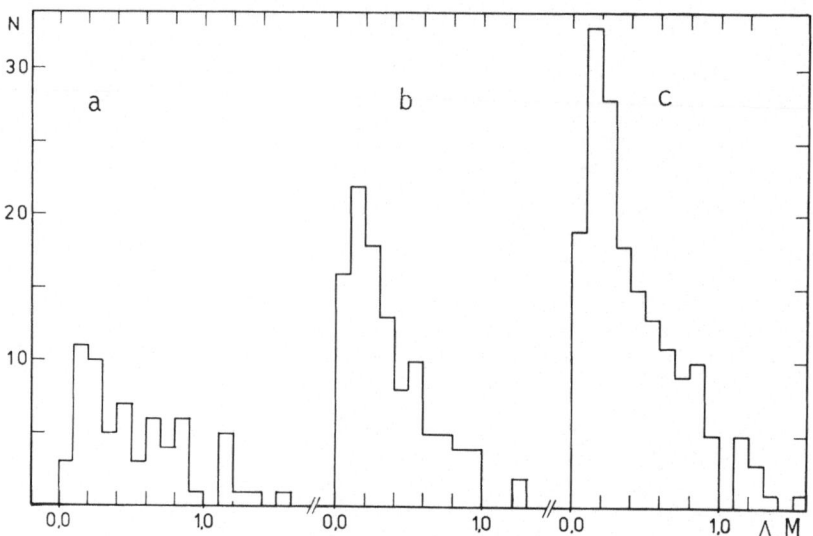

Fig. 3.   The distribution of the deviation ΔM from the Main Sequence for (a) hot stars $B - V < 0.3$, (b) late-type stars $B - V > 0.3$, and (c) for all used stars.

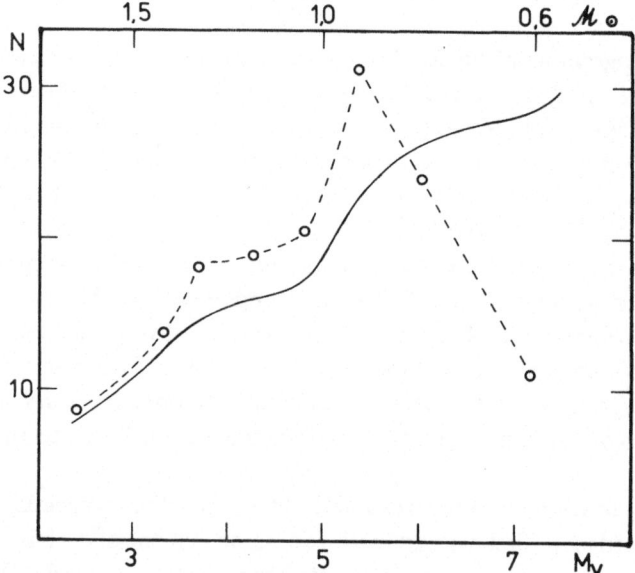

Fig. 4.   Smoothed van Rhijn function for our star mixture (solid line) and the calculated mean luminosity function (broken line).

for $|\Delta M| = 0\overset{m}{.}0$. We have calculated the luminosity function for the unseen companion for three different stellar types of the component A corresponding to the colors $B - V = -0\overset{m}{.}10, 0\overset{m}{.}25$, and $0\overset{m}{.}60$. These, in turn, correspond to stars of 3.8, 1.66, and

1.10 solar masses, respectively. Figure 4 shows the smoothed van Rhijn function (solid line) and the calculated mean luminosity function (giving double weight to the $B - V = 0^{m}.25$ model). The effects of the third companion are included as long as their influence on the integrated brightness and color of the component has significant value. As long as the mass ratio $q = \mathfrak{M}_B/\mathfrak{M}_A$ is not smaller than 0.5, the distribution is indistinguishable to the van Rhijn luminosity function. For smaller tertiary masses the number of unseen companions decreases rapidly. In Figure 5 we show the highest

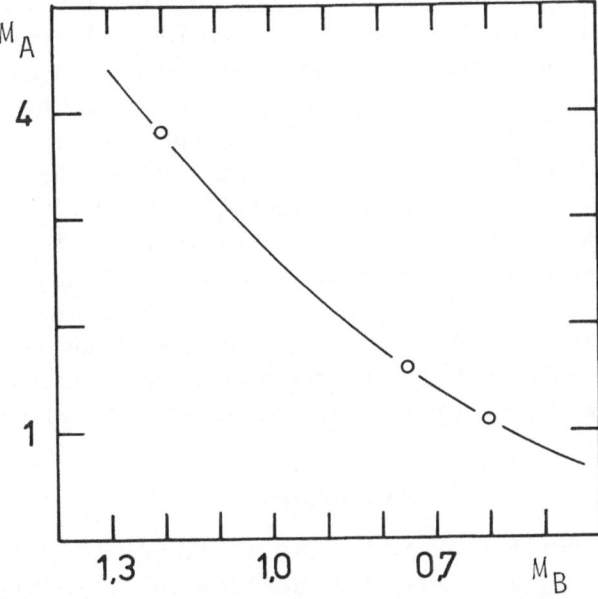

Fig. 5.   Turning points in the luminosity function for the masses of the third companions $\mathfrak{M}_B$ as a function of the mass of correspondent component $\mathfrak{M}_A = 3.8$, 1.66, and 1.1 solar masses.

probability for the masses of third companions for the given mass of the component calculated for our three models. This result is in very good agreement with observations of Abt and Levy (1976, 1978). According to their calculations for binaries with periods less than 100 years, the number of companions with 1.2 solar masses corresponds to the van Rhijn function. The number of companions of only 0.6 solar masses drops significantly in comparison to this function. More realistic discussions will have to include effects of observational selections and incompleteness of material. Looking again to the histogram 3c, we are able to attribute the intrinsic scatter of the Main Sequence to the presence of a large number of triple systems among the so-called binary stars. Less than 13% of our 'binaries' are probably simple binary systems. However, the masses of the components in a multiple system are of the same order of magnitude.

    Detailed investigations of F and G dwarfs and subgiants by Abt and Levy (1976, 1978) also implied that single stars are rare. The discoveries of spectroscopic binaries are still incomplete even in the solar neighborhood. Recently, also Fofi et al. (1983)

reported the existence of a group of binary systems with high eccentricity and inter-mediate period with an unusually high frequency of triple systems. Our results are not in agreement with the calculations of Lucy and Ricco (1979). They have proposed a distribution sharply peaked at $q = 1$ ($|\Delta M| = 0\overset{m}{.}75$) for short period binaries. Part of the controversy is due to the complex selection effects hidden in the catalogues and the restriction to very short period binaries investigated by Lucy and Ricco. Kuiper has already obtained a distribution function from a sample of bright Main-Sequence stars that is essentially constant, but with some decrease at the low mass values. Jaschek (1976) and Bettis (1975) have analysed color magnitude diagrams of clusters to obtain the deviation above the Main Sequence of the binaries. The results are controversial to Lucy and Ricco. Jaschek's results show that the distribution fucntion is decreasing function of the mass ratio – i.e., less massive companions are much more frequent than massive ones. Finally, Trimble (1978) has published an updated version of her investi-gation of the distribution of mass ratios in spectroscopic binaries. She claims a bimodal distribution with peaks near $q = 0.3$–0.3 and 0.9–1.0. To account for this distribution it has been suggested that two different mechanisms are involved in the formation of binaries, with capture being the predominant process for wide binaries and fission the relevant process for close binaries. The cut-off at small values of q can be attributed largerly to observational selection, as the radial velocity amplitude is roughly propor-tional to q for small q. This means that by changing the incompleteness correction, the observed peak for $q = 0.2$–0.3 can be shifted to 0.5 value proposed in this paper.

Acknowledgements

I gratefully acknowledge Dr W. Gliese and Dr J. C. Mermilliod for helpful comments and Dr C. Pilcher for improvements of the English version of the text. This investigation is supported by the Austrian Science Foundation 'Fonds zur Förderung der wissen-schaftlichen Forschung', No. P5073.

References

Abt, H. A. and Levy, S. G.: 1976, *Astrophys. J. Suppl.* **30**, 273.
Abt, H. A. and Levy, S. G.: 1978, *Astrophys. J. Suppl.* **36**, 241.
Bettis, C.: 1975, *Publ. Astron. Soc. Pacific* **87**, 707.
Eggen, O. G.: 1963, *Astron. J.* **68**, 483.
Eggen, O. G.: 1965, *Astron. J.* **70**, 19.
Fofi, M., Maceroni, C., Maravalle, M., and Paolicci, P.: 1983, *Astron. Astrophys.* **124**, 313.
Hurly, P. R. and Warner, B.: 1982, *Area Scanner Observations of Close Visual Double Stars II: Results for 153 Southern Stars* (preprint).
Jaschek, C.: 1976, *Astron. Astrophys.* **50**, 185.
Lucy, L. B. and Ricco, E.: 1979, *Astron. J.* **84**, 401.
Lutz, T. E.: 1971, *Publ. Astron. Soc. Pacific* **83**, 488.
Murphy, R. E.: 1969, *Astron. J.* **74**, 1082.
Rakos, K. D., Albrecht, R., Jenkner, H., Kreidl, T., Michalke, R., Oberlerchner, D., Santos, E., Scher-mann, A., Schnell, A., and Weiss, W. W.: 1982, *Astron. Astrophys. Suppl.* **47**, 221.
Stephenson, C. B.: 1960, *Astron. J.* **65**, 60.
Trimble, V. L.: 1978, *Observatory* **98**, 1₆₂

THE FREQUENCY OF BINARIES WITH A DEGENERATE COMPONENT*

J. L. HALBWACHS

Observatoire de Strasbourg, France

(Received 18 September, 1984)

Abstract. The frequency of binaries with degenerate secondary components was evaluated according to the spectral types of the primaries. It appears that this proportion is 25% for binaries with giant primary components, and less than about 17% for dwarfs.

1. Introduction

The binaries usually observed are constituted of dwarf or of giant stars. In some cases however, one of the components may be more evolved and is at present a faint degenerate object, white dwarf, neutron star or black hole. Such a binary is very difficult to detect and is often considered as a single star. Moreover, since low-mass stars require more time to evolve, the initially secondary components appear as the brighter components after the primaries have reached their final stage. In other words, observers often see a single star with initial mass $\mathfrak{M}_2$ instead of a binary system with an initial primary mass $\mathfrak{M}_1$. This is important for some statistical studies – like the estimation of the frequency of binaries among stars, or the study of the luminosity function. It seems then important to provide a theoretical estimation of the frequency of these particular binary systems.

2. Definitions

Let us consider a sample of binaries at different stages of their evolution. For a given range in spectral type of the currently brighter components, we want to estimate the proportion of systems with a degenerate companion; let 'R' be this proportion. In practice, it is also useful to consider, for a given range of current primaries, the ratio between systems for which the initial primaries are always the brighter components and systems for which the current primaries were initially the secondary components, the initial primaries being evolved into degenerate objects; let 'r' be this ratio. R may be derived from r by the relation

$$R = r/(1 + r). \tag{1}$$

* Communication presented at the International Conference on 'Astrometric Binaries', held on 13–15 June, 1984, at the Remeis-Sternwarte Bamberg, Germany, to commemorate the 200th anniversary of the birth of Friedrich Wilhelm Bessel (1784–1846).

For stars with current primaries with initial masses $\mathfrak{M}$, r is given by

$$r(\mathfrak{M}) = \frac{\displaystyle\int_{q=0}^{q_M} \int_{t=\min(T_G, T(\mathfrak{M}))}^{\min(T_G, T(\mathfrak{M}/q))} \frac{1}{q} \Psi(\mathfrak{M}/q) f(q) B(T_G - t)\, dt\, dq}{\displaystyle\int_{0}^{\min(T_G, T(\mathfrak{M}))} \Psi(\mathfrak{M}) B(T_G - t)\, dt}, \tag{2}$$

where $B(t)$ is the stellar birthrate ($t = 0$ at the epoch of the origin of the galactic disk); $\Psi(\mathfrak{M})$ is the initial mass function; according to Salpeter (1955),

$$\Psi(\mathfrak{M}) = \mathfrak{M}^{-2.35}; \tag{3}$$

Miller and Scalo (1979) have recently confirmed the validity of this law for stars with masses in the range of 0.5 to 10 solar masses; $f(q)$ is the distribution of binaries according to their mass ratios, $q = \mathfrak{M}_2/\mathfrak{M}_1$; $T(\mathfrak{M})$ is the time that a star with initial mass $\mathfrak{M}$ spend in the dwarf and in the giant stages; according to Neckel (1975) and with the results of Mengel *et al.* (1979) one has

$$T(\mathfrak{M}) = (10\,\mathfrak{M}^{-3.25} + 1.5\,\mathfrak{M}^{-2.75}) \times 10^9 \text{ yr}, \tag{4}$$

(the two terms in Equation (4) are, respectively, the durations of the Main-Sequence and giant stages); T_G is the age of the galactic disk; following Neckel (1975), we assume that $T_G = 10 \times 10^9$ yr; q_M is 1 for stars such that $T(\mathfrak{M}) < T_G$ and else $q_M = \mathfrak{M}_{\text{lim}}/\mathfrak{M}$, $\mathfrak{M}_{\text{lim}}$ being the initial mass of the stars with longevity equal to the age of the galactic disk; with the hypotheses assumed, $\mathfrak{M}_{\text{lim}} = 1.045\,\mathfrak{M}_\odot$.

3. Calculation of r and R

3.1. DEPENDENCE OF r UPON THE MASS RATIOS

In order to solve Equation (2), we can use an approximation for $T(\mathfrak{M})$; instead of Equation (4), we assume that

$$T(\mathfrak{M}) \propto \mathfrak{M}^{-3.25}. \tag{5}$$

On the other hand, we consider two different cases for $B(t)$:

(a) $B(t) = C^{st}$, $\qquad\qquad\qquad\qquad\qquad\qquad\qquad\qquad\qquad\qquad$ (6)

and

(b) $B(t) = 2 \times (1 - t/T_G)/T_G$; $\qquad\qquad\qquad\qquad\qquad\qquad\qquad$ (7)

The case (b) corresponds to a linear stellar birthrate decreasing since the origin of the Galaxy and null at present.

For a sample of binaries with the same initial mass ratio, q, and with primary masses larger than $\mathfrak{M}_{\text{lim}}$, we obtain in the case (a)

$$r(q) = q^{1.35} - q^{4.60} \tag{8}$$

and in case (b)

$$r(q) = q^{1.35} - q^{7.85} . \tag{9}$$

The proportion of binaries with a degenerate component, R, may now be derived from Equations (8), (9), and (1) and is plotted in Figure 1.

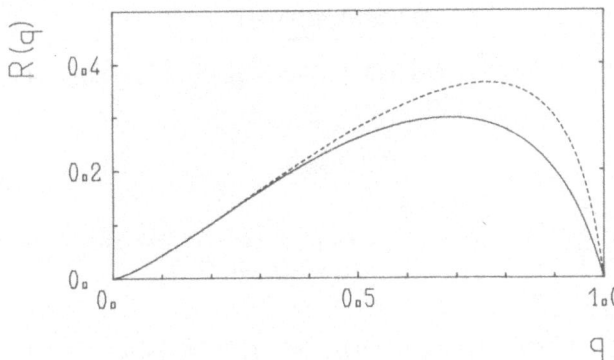

Fig. 1.   The proportion of binaries with one degenerate component R, related to the mass ratios q; full line: case of a constant stellar birthrate; dotted line: linear decreasing stellar birthrate reaching zero at present.

It appears that the frequency of binaries with a degenerate component is greater when the distribution of binaries among mass ratios contains many systems with q near 0.7; nevertheless, this frequency cannot be more than about 40%.

3.2. THE PROPORTION OF BINARIES WITH ONE DEGENERATED COMPONENT RELATED TO THE SPECTRAL TYPES OF THE CURRENT PRIMARIES

Let us now assume that binaries follow the distribution of mass ratios derived by Halbwachs (1983) from the bright visual binaries and assume also that the stellar birthrate is constant. It is then possible to compute R from Equations (2) and (1), or, as it was done in fact, by a sampling experiment taking into account all the hypotheses. Table I provides the results; since we wanted the results in terms of spectral type rather

TABLE I

Spectral types of the current primaries	Proportions of systems with one degenerate component
A5–F0 V	17%
F0–G2 V	17%
G2–K0 V	15%
K0–M0 V	10%
Late-type giants	25%

than in terms of masses, we made the conversion by means of the relation given by Schmidt-Kaler (1982).

4. Conclusion

The proportion of binaries with one degenerate component seems to be about 25% in systems with giant primaries and 17% in systems with Main-Sequence early-type primaries.

Acknowledgement

It is a pleasure to thank Prof. C. Jaschek for his support and guidance in this project.

References

Halbwachs, J. L.: 1983, *Astron. Astrophys.* **128**, 399.
Mengel J. G., Sweigart, A. V., Demarque, P., and Gross, P. G.: 1979, *Astrophys. J. Suppl. Ser.* **40**, 733.
Miller, G. E. and Scalo, J. M.: 1979, *Astrophys. J. Suppl. Ser.* **41**, 513.
Neckel, H.: 1975, *Astron. Astrophys.* **42**, 379.
Salpeter, E. E.: 1955, *Astrophys. J.* **121**, 161.
Schmidt-Kaler, T.: 1982, in *Landolt Bornstein*, Springer Verlag, Berlin, VI2b.

TRAPEZIUM-TYPE MULTIPLE SYSTEMS*

L. V. MIRZOYAN

Byurakan Astrophysical Observatory, Armenia, U.S.S.R.

and

G. N. SALUKVADZE

Abastumani Astrophysical Observatory, Georgia, U.S.S.R.

(Received 25 June, 1984)

Abstract. The non-stable nature of stellar motions in Trapezium-type multiple systems is discussed. The disintegration time of such systems is several orders of magnitude less than the life-time of their components, and considerably less than the disintegration time of stellar associations. The importance of ground-based and orbital observations of Trapezium-type systems is stressed.

1. Introduction

In double star systems, both components are rotating around the center of gravity by Kepler's laws. These motions are periodical and can go on for a very long time. Therefore, such systems can be considered as dynamically stable. In stellar systems with a larger number of components such a structure is usually observed (triple star = double star + comparatively distant component; star quartet = double star + double star with a distance much larger than the mutual distances of the components of double stars etc.) which is connected with periodic (Keplerian or quasi-Keplerian) motions. It is evident, that such systems can also be considered as dynamically stable. All systems of the mentioned type are called multiple systems of ordinary type (Ambartsumian, 1954).

There are, however, stellar systems for which the structure differs from the structure of the ordinary type multiple systems in that they contain at least three components, the mutual space distances between which are of the same order of magnitude (differ by not more than, for example, a factor of three). Due to such a structure the motions of stars in these systems are non-periodic; they cannot be stable and have to disintegrate. The famous Trapezium in Orion is a system of this type and multiple systems of such a type were named Trapezium-type multiple systems after it (Ambartsumian, 1954).

In this paper some results of the study of the Trapezium-type multiple systems are given.

* Communication presented at the International Conference on 'Astrometric Binaries', held on 13–15 June, 1984, at the Remeis-Sternwarte Bamberg, Germany, to commemorate the 200th anniversary of the birth of Friedrich Wilhelm Bessel (1784–1846).

2. Non-Stable Nature of Motions in Trapezium-Type Multiple Systems

Ambartsumian (1954) has shown, that the nature of star motions in the Trapezium-type multiple systems must substantially differ from the motions in ordinary type multiple systems.

As has been mentioned the motions of stars in the ordinary type systems are Keplerian or quasi-Keplerian. In this respect they differ greatly from star clusters where due to close passings of stars the exchange of kinetic energies takes place. As a result, irregular forces cause in time the gradual disintegration of the cluster (Ambartsumian, 1938).

Due to stellar motions in them, the Trapezium-type multiple systems behave like star clusters. The only difference is the total number of components. Therefore, in order to estimate the relaxation time, T for Trapezium-type systems, one can use the formulae derived by Ambartsumian for star clusters (see, for example, Chandrasekhar, 1942):

$$T = 8.8 \times 10^5 \sqrt{\frac{NR^3}{m}} \frac{1}{\log N - 0.45} \text{ yr},$$

where N is the number of stars in the system, R the radius of system in parsecs, and m the mean mass of stars.

Using this formula with N equal to some units, R of the order of 10 000 AU, and m of the order of the solar mass a value of the order of 2×10^6 yr has been obtained (Ambartsumian, 1954) for the disintegration time of a Trapezium-type multiple system.

This means that a Trapezium-type multiple system has enough time to disintegrate while each star-component in it performs only some revolutions around the centre of gravity of the system.

This estimation concerns Trapezium-type systems with negative total energy. General considerations give every reason, however, to assume that many Trapezium-type systems can possess a positive total energy. In this case the disintegration time of the Trapezium-type system may be only 10^5 yr and less (Ambartsumian, 1954).

Current data confirm this conclusions on the non-stable nature of motions in Trapezium-type systems and on the dynamical instability of these systems.

This question is considered in detail by Mirzoyan and Salukvadze (1984). Here we shall consider only two results which support this conclusion.

Kinematics of the Trapezium-type systems have been studied in papers by one of the authors (see, for example, Salukvadze, 1984). This study is based on the observations of 38 Trapezium-type multiple systems, the main members of which belong to O-B2 spectral classes. The results of relative position measurements of their components have been used. The data were taken from different catalogues of double stars, card-catalogues of double stars of the Nice (France) and Naval (USA) Observatories, as well as from measurements of photographic observations obtained at Abastumani Astrophysical Observatory.

Among all 38 Trapezium-type multiple systems, the data are sufficient by only 15 systems: ADS 719, 2783, 2843, 3709, 4241, 4728, 5322, 5977, 13 374, 13 626, 14 526,

14 831, 15 184, 16 095, and 16 381, the majority of their components were observed for at least five times.

The observational material concerning the mutual distances of the components of these 15 systems, includes in most cases a time interval of more than 100 yr.

Based on this astrometric material, the graphs of the dependence: mutual distance of components – time (epoch) of observation have been drawn. They indicate an expansion of 14 Trapezium-type systems out of 15 studied. As an example Figure 1 shows the graphs for ADS 719.

For the problem considered here, the paper of Allen and Poveda (1974) on the study of the dynamical evolution of Trapezium-type systems is of interest. It is based on the numerical integration of the motion equations of these system components.

In this paper the motions of stars in Trapezium-type multiple systems were studied by assuming that the total energy of the systems is negative. Each of them consists of

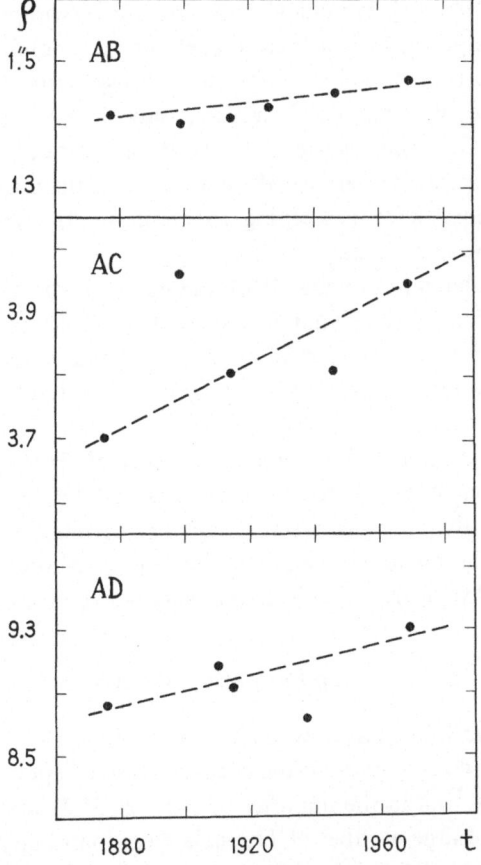

Fig. 1. Observed dependences (ρ, t) showing the expansion of the Trapezium-type system ADS 719 (Salukvadze, 1984), where ρ (in arc sec) is the angular distance from the main star and t is the time (epoch).

six components which have different parameters of structure, inner spheres with a radius of 5000 AU. The masses of the components are 50, 20, and 15 $M_\odot$.

The results of this paper by Allen and Poveda (1974) show that after 10^6 yr of dynamical evolution, $\frac{2}{3}$ of all systems considered still continue to remain as Trapezium-type systems. In this result Allen and Poveda (1974) found a contradiction to the idea of the dynamical instability of the Trapezia. It turns out, however, that this conclusion was based on an incorrect interpretation of the results obtained.

Indeed, as it has been shown by Mirzoyan and Mnatsakanian (1975), this result means that for a Trapezium-type system the probability to keep its natural configuration during 10^6 yr is equal to $\frac{2}{3}$. Therefore, already after 2×10^6 yr of dynamical evolution more than half, exactly $1 - (\frac{2}{3})^2 = \frac{5}{9}$, of the studied sample of Trapezium-type systems will lose their Trapezium configuration. In other words, the time of semi-disintegration for the Trapezium-type multiple systems with negative total energy, is less than 2×10^6 yr.

The results of Allen and Poveda (1974) concerning the structure of studied Trapezium-type systems after 10^6 yr of dynamical evolution are very impressive. They show, that out of 30 studied Trapezia, 11 lost their Trapezium-type configuration: 3 systems disintegrated leaving double stars and 8 transformed into systems of ordinary type. Out of 19 other Trapezia keeping their configuration, only in 6 systems the number of components is unchanged, while 6 systems lost one of their components each, and 7 systems lost two components. Otherwise, these 13 systems disintegrated partly during 10^6 yr. At last 6 Trapezium systems keeping all components have undergone some changes: 5 of them extended in sizes.

Hence, the results of Allen and Poveda (1974) are new and strong evidence in favour of the very important idea of the dynamical instability of Trapezium-type multiple systems, which during a time of the order of 10^6 yr or less, must disintegrate completely or partly lose some components and be transformed into ordinary type systems with a smaller number of members.

The results of observational and theoretical studies of Trapezium-type multiple systems give every reason to conclude that these systems having unusual structural space configurations are dynamically instable and are disintegrated at present. The disintegration time for a Trapezium depends on the sign of its total energy. It amounts to approximately 2×10^6 yr, if the system total energy is negative and 10^5 yr and less, if it is positive.

3. On Group Formation of Stars

The statistical study of the Trapezia showed (Ambartsumian, 1954) that these multiple systems contain as main components in most cases stars of spectral classes of O-B.

Further studies showed that in the majority of cases real* Trapezia are members of known OB-associations. Large number of Trapezia were found in T associations (see, for example, Mirzoyan and Salukvodze, 1984; Salukvadze, 1982).

* Some multiple systems not having Trapezium-type configuration can be observed in the sky as Trapezia due to their projection ('pseudo-Trapezia').

Taking into account that the components of a physical system have a common origin (Ambartsumian, 1947, 1954), one can conclude that all members of a Trapezium-type system were formed together. The existence of a large number of Trapezium-type multiple systems in the regions of star formation is therefore telling evidence in favour of the idea of group formation of stars (Salukvadze, 1982; Ambartsumian, 1947,1954; Mirzoyan, 1982).

At the same time, the existence of real Trapezia, systems of recently formed stars, indicates an important feature of star formation process: the stars are formed in stellar associations in groups; together with dynamically stable groups (double stars, ordinary type multiple systems, many star clusters), dynamical unstable groups (Trapezium-type multiple systems, associations and, probably, some star clusters) are formed.

4. Conclusions

The discovery of the existence of stellar associations – i.e., non-stable stellar systems (Ambartsumian, 1949) where the star formation process is continuing at present – has played a fundamental role in establishing new ideas on the origin and evolution of stars and stellar systems. Due to this discovery, a possibility appeared for the first time to study phenomena which are connected with the star formation process and which are based directly on astronomical observations (Ambartsumian, 1947; Mirzoyan, 1982).

Important results have been obtained in this field after the separation and study of multiple systems of new type-Trapezium type systems. It turned out, that these systems are usually members of stellar associations, are dynamically unstable, and consist of very young stars and are presently disintegrating. The disintegration of Trapezium-type systems time is many orders of magnitude less than the life-time of their components, and considerably less than the disintegration time of stellar associations. After the disintegration of Trapezia, their members gradually enrich, therefore, the general stellar field of the galaxy. The existence of Trapezium-type multiple stars is new telling evidence in favour of the idea of group formation of stars.

In conclusion let us mention some problems of the study of Trapezia which seem to be most important.

(1) The search of new Trapezium-type multiple systems containing OB stars.

(2) The measurements of proper motions and radial velocities of stars in Trapezia, which can be considered as real systems of this type (Trapezia connected with OB stars).

(3) The physical study of the components in Trapezia containing OB stars.

(4) The physical and statistical studies of Trapezia in T-associations.

For the solution of these problems, ground based and orbital astronomical observations of Trapezia are very important. Orbital observations must be particulary effective for measurements of the proper motions in Trapezia.

These investigations will contribute to the study of the earliest stages of stars, directly following their formation. They can also be effective for the study of dynamical pecularities of Trapezia, in particular, for the ascertainment of the existence of stellar systems having positive total energy in the galaxy.

References

Allen, C. and Poveda, A.: 1974, in Y. Kozai (ed.), 'The Stability of the Solar System and Small Stellar Systems', *IAU Symp.* **62**, 239.

Ambartsumian, V. A.: 1938, *Ann. Leningrad State University* **22**, 19.

Ambartsumian, V. A.: 1947, 'Stellar Evolution and Astrophysics', *Ac. Sci. Armenian SSR*, Yerevan.

Ambartsumian, V. A.: 1949, *USSR Astron. J.* **26**, 3.

Ambartsumian, V. A.: 1954, *Transactions IAU*, Vol. VIII, University Press, Cambridge, p. 665.

Ambartsumian, V. A.: 1954, *Comm. Byurakan Obs.* **15**, 3.

Chandrasekhar, S.: 1942, *Principles of Stellar Dynamics*, University of Chicago Press, Chicago.

Mirzoyan, L. V. and Mnatsakanian, M. A.: 1975, *Astrofizika* **11**, 551.

Mirzoyan, L. V.: 1982, in Z. Kopal and J. Rahe (eds.), 'Binary and Multiple Stars as Tracers of Stellar Evolution', *IAU Colloq.* **69**, 61.

Mirzoyan, L. V. and Salukvadze, G. N.: 1984, *Astrofizika* (in press).

Salukvadze, G. N.: 1982, in Z. Kopal and J. Rahe (eds.), 'Binary and Multiple Stars as Tracers of Stellar Evolution', *IAU Colloq.* **69**, 109.

Salukvadze, G. N.: 1984, *Astrofizika* (in press).

A SEARCH FOR CPM STARS*

J. L. HALBWACHS

Observatoire de Strasbourg, France

(Received 16 July, 1984)

Because of the interest that common proper motion stars offer both for theoretical studies and for observations, I have searched common proper motion pairs in the AGK3 catalog. About 300 pairs were found. Full results will be published in a near future.

* Communication presented at the International Conference on 'Astrometric Binaries', held on 13–15 June, 1984, at the Remeis-Sternwarte Bamberg, Germany, to commemorate the 200th anniversary of the birth of Friedrich Wilhelm Bessel (1784–1846).

THE SPECTROSCOPIC STUDY OF ASTROMETRIC BINARIES*

GORO ISHIDA

Tokyo Astronomical Observatory, University of Tokyo, Japan

(Received 13 June, 1984)

Abstract. The spectroscopic observations of several astrometric binaries, obtained by the author during the past 20 years, are described. Main emphasis is placed on the determination of orbital elements of visual binaries, and the detection of unseen companions.

(1) I have continued spectroscopic observations of several astrometric binaries during the last twenty years. The dispersion power of the spectrograph at the coudé focus of the 74 inch reflector at Okayama Station of the Tokyo Astronomical Observatory is 4 Å mm^{-1}. The measuring machine is a so-called Hartmann type comparator made by the NIKON optical company. The accuracy of the measurement of the radial velocities is less than ± 1 km.

(2) The first purpose of my study is to determine the sign of the inclination value of the orbits of visual binaries, which can never be known by astrometric methods. It is very important for the statistical study of the distribution of the orbital planes. Some examples are shown as follows:

(a) 1 Geminorum: Kui 23; 5h58.0 m + 23 : 16; 4.7 (G8III) − 5.1; orbital elements:

$$P = 13.17 \text{ yr}, \quad a = 0.19'', \quad i = 57.3°, \quad e = 0.325, \quad \omega = 192° .$$

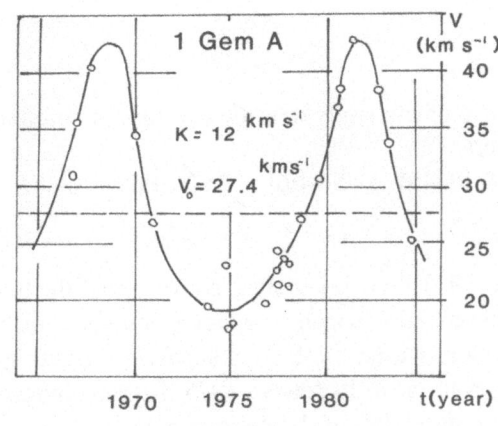

Fig. 1.

* Communication presented at the International Conference on 'Astrometric Binaries', held on 13–15 June, 1984, at the Remeis-Sternwarte Bamberg, Germany, to commemorate the 200th anniversary of the birth of Friedrich Wilhelm Bessel (1784–1846).

Astrophysics and Space Science **110** (1985) 161–162. 0004–640X/85.15
© 1985 *by D. Reidel Publishing Company.*

I took 24 spectrograms between 1967 to 1983. The observed values are plotted on the figure. These values coincide very well with the velocity curve drawn with the above elements with the equation

$$V = V_0 + K\{e \cos \omega + \cos(v + \omega)\} ;$$

where it is assumed that $V_0 = 27.4$ km s^{-1} and $K = 12$ km s^{-1}.

The O–C value is -1.1 km s^{-1}.

(b) Epsilon Hydrae (ADS 6993) 8h41.5 m + 06:47; 3.8 (G0III) – 4.7; orbital elements:

$$P = 15.05 \text{ yr}, \quad a = 0.226'', \quad i = 50.4°, \quad e = 0.67, \quad \omega = 264° ;$$

we can calculate the velocity curve assuming that $V_0 = 36$ km s^{-1} and $K = 9$ km s^{-1}.

In this case 21 observed points taken in 1968–1983 show a nearly good fitting. But a preliminary calculation shows a somewhat higher eccentricity value of about $e = 0.72$.

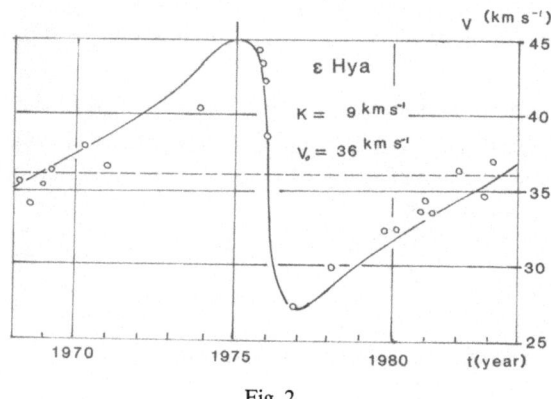

Fig. 2.

(3) The main purpose of my study is to detect unseen companions by the velocity variations of the objects.

Mu Draconis (ADS 10345); 5.65(dF6) – 5.70(dF6):

$$P = 480 \text{ yr}; \quad a = 3.33''; \quad i = 143.4; \quad e = 0.375; \quad \text{and} \quad \omega = 204°3 .$$

Dr Strand stated in 1943 that this system shows small fluctuations of the periode 3.2 years; but we can never certify which component actually shows the small fluctuation in the main orbital motion. I used the Lick collection of spectrograms, and continued the spectroscopic observations from 1966 to 1978. I concluded that the secondary star might have an unseen companion of the periode 6.22 years.

(4) The astrometric orbit is the projection onto the $x - y$-plane tangential to the celestial sphere. The radial velocity variation relates to the independent z-axis. Since it might expose another new face of astrometric binaries, their spectroscopic observation is strongly recommended.

G 82–23: A NEW SUBDWARF-WHITE DWARF BINARY

I. BUES

Dr. Remeis-Sternwarte, Bamberg

and

G. RUPPRECHT

Max-Planck-Institut für Plasmaphysik, München

(Received 14 January, 1985)

Abstract. Photometry and spectrophotometry of the proper motion star G 82–23 are presented. A comparison with subdwarfs and white dwarfs in the same range of temperature shows only partial agreement. If the parallax is taken into account, the best explanation of this object seems to be a binary structure with a K-subdwarf and a DC-white dwarf.

1. Introduction

In order to investigate the apparent deficit of white dwarfs in the solar neighbourhood as seen from the Gliese catalogue (1969) and discussed by Liebert *et al.* (1979) we started a photometric programme for candidates of white dwarf suspects from the Lowell Observatory GD and G lists (Giclas *et al.*, 1978, 1980).

During these observations of up to now 230 stars two cool objects were included which by all means of combined photometric diagrams fitted a white dwarf solution with a photometric M_V of $14\overset{m}{.}5$ or $14\overset{m}{.}2$, respectively. The spectra, however, revealed features not typical for white dwarfs in the appropriate range of effective temperature.

Here we would like to discuss one star, G 82–23 = L 951–48 ($V = 14\overset{m}{.}7 - 14\overset{m}{.}8$, $\alpha(1950) = 4^h27^m2^s$, $\delta(1950) = -30°09.6$) only, for which we obtained additional observational material. A new parallax has been provided by van Altena (1984). The solution for spectra and colours is the assumption of a binary system with a cool subdwarf as the main component for the spectral features for wavelengths > 420 nm and a white dwarf of $T_{\mathrm{eff}} > 8000$ K for the blue part of the spectrum and blue colours.

2. Observations and Data Analysis

During three stays at the European Southern Observatory (La Silla, Chile) in November 1981, 1983 and October 1982 photometric observations at the 1 m-telescope and image tube spectra at the 1.52 m-telescope were taken by one of us (I.B.). Our method of data evaluation by combined two-colour diagrams is explained by Rupprecht (1983) and Rupprecht and Bues (1983) in detail.

---

* Communication presented at the International Conference on 'Astrometric Binaries', held on 13–15 June, 1984, at the Remeis-Sternwarte Bamberg, Germany, to commemorate the 200th anniversary of the birth of Friedrich Wilhelm Bessel (1784–1846).

Astrophysics and Space Science **110** (1985) 163–168. 0004–640X/85.15

© 1985 *by D. Reidel Publishing Company.*

Here a short description will be sufficient for the understanding of our determination of the photometric classification of G 82–23: If possible, at the same time, *UBV*-, Strömgren-, and *VRI*-photometry should be obtained for catalogue stars with $11.5 < m_v < 15.0$ with the same standards to provide homogeneous material for various combined two-colour diagrams. A first classification is possible by use of 'classical' two-colour diagrams $(U - B)/(B - V)$ and $(u - b)/(b - y)$: Main-Sequence stars can be easily separated from white dwarfs of spectral type DA and subdwarfs in the range of $12\,000 > T_{\text{eff}} > 6000$ K. For hot stars, the $(u - b)/(U - V)$ diagram allows an identification of white dwarfs as demonstrated by Rupprecht and Bues (1983). For cool objects the $(R - I)/(U - V)$ or $(R - I)/(u - b)$ diagrams are more appropriate to decide on white dwarfs of types DA and DC, subdwarfs and Main-Sequence stars.

This diagram is shown in Figure 1, where the position of G 82–23 is marked by a special symbol. All white dwarfs occupy the region around the black-body line with the hydrogen-rich DA stars above it and the DCs below.

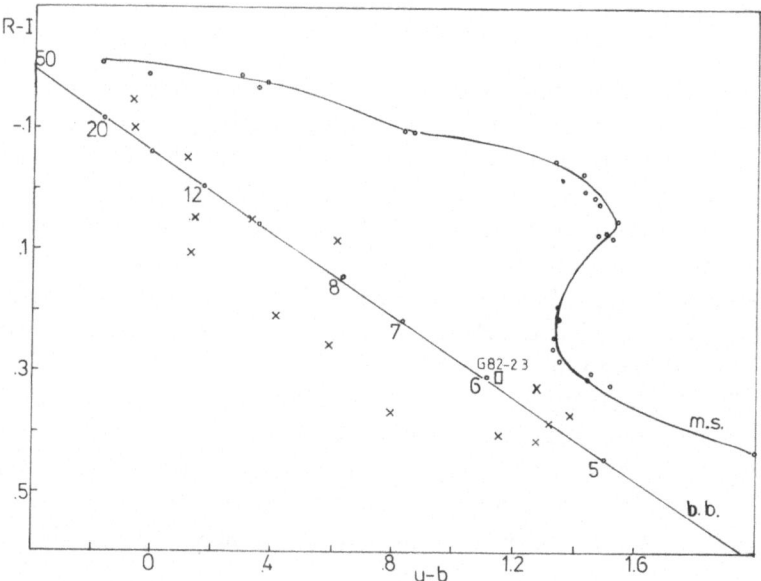

Fig. 1.   Two-colour diagram $(R - I)/(u - b)$. $\times$ = new white dwarfs with additional information from spectra. $\square$ = G 82–23 (obs. 1981).

G 82–23 is situated at the narrow region between Main Sequence and black body, yet the accuracy of a single measurement (1981 data) is $0\rlap{.}''05$ in $(u - b)$ and $0\rlap{.}''06$ in $(R - I)$.

Table I contains our photometric data of G 82–23 – together with our data of some subdwarfs and white dwarfs in the same range of at least *one* colour as well as colours of G 82–23 obtained by Lacombe and Fontaine (1981), Harrington and Dahn (1980),

TABLE I

Photometric data of G 82–83 compared to extremely metal-poor subdwarfs and white dwarfs

	y	$b - y$	m_1	c_1	$u - b$
G 82–23	14.81	0.42	0.19	0.05	1.26 (1983)
	14.75	0.54	− 0.04	0.14	1.14 (1981)
	14.73	0.50	0.10		1.39 (L)
GD 801	13.79	0.46	0.02	0.30	1.26
GD 1347	14.12	0.43	− 0.01	0.29	1.13
GD 806	14.98	0.43	0.16	− 0.03	1.15
G 82–44	14.13	0.33	0.17	0.12	1.12
GH 7–21	13.44	0.46	0.05	0.09	1.11
G 14–24	13.33	0.51			1.21 (W)
		0.54			1.36 (B)

	V	$B - V$	$U - B$	$R - I$	
G 82–23	14.70	0.71	− 0.22	0.31	(1981)
	14.74	0.66	− 0.12		(H)
	14.70	0.72	− 0.15		(E)
GD 801	13.80	0.60	− 0.06	0.40	
GD 1347	14.23	0.49	− 0.12	0.32	
G 14–24	12.82	0.72	0.04		(H)

(L) = Lacombe and Fontaine (1981), (W) = Wegner (1983), (H) = Harrington and Dahn (1980), (E) = Eggen and Greenstein (1965), (B) = Bessell and Wickramasinghe (1979).

Eggen and Greenstein (1965), and observations of GH 7–21 as published by Wegner (1983).

The table of the Strömgren colours clearly shows the difference between G 82–23 and the other stars in $(u - b)$. Our object is − even if we take care of the variability in m_1 and $u - b$ − the bluest star in this range of $(b - y)$. The change in colour would move G 82–23 in the normal two-colour diagrams from a position above the black-body line to one below, but far away from metal-poor subdwarfs of high velocity. In Figure 1 only the value by Lacombe and Fontaine shifts the position close to the Main Sequence, the other values give a better fit with the black-body line.

So our first guess was a cool hydrogen-rich white dwarf just as mentioned by Eggen and Greenstein (1965) with spectral-type DFs.

As for all our classified objects various colour-luminosity relations were used to derive a photometric parallax. The absolute magnitude M_V was determined from the $M_V/(B - V)$ relation given by Greenstein (1976), the $M_V/(b - y)$ relation by Greenstein (1984) and from linear regressions $M_V/(u - b)$ and $M_V/(R - I)$ which we calculated for stars with known trigonometric parallaxes. The result for the data of 1981 is $\langle M_V \rangle = 14^m\!.5 \pm 0^m\!.5$.

The resulting distance would be 11 pc, which should be measurable for the Flagstaff-parallax programme. Yet the relative parallax (published by Harrington and Dahn,

1980) yielded a negative value and their absolute value with $\pi = 0\overset{''}{.}0013$ placed the star in their $M_V/(B - V)$ diagram to a Main-Sequence position (G6V) in contradiction to all photometric determinations and the spectral classification by Eggen and Greenstein (1965). In addition, the tangential velocity would be ~ 1000 km s^{-1}, abnormally high for an object of this class. A new parallax determination by Van Altena (1984) with $\pi = 0\overset{''}{.}0033$ results in a more likely tangential velocity of ~ 300 km s^{-1} for subdwarfs, the colours, however, remain far too blue.

Our information from two spectra (dispersion 171 Å mm^{-1}) confirm the assumption that the hydrogen profiles of the star belong to a subluminous object, but there are strong features of CH and very weak Fe I lines. Figure 2 shows one spectrum compared to the white dwarf GD 806 and the extremely metal-poor subdwarf GD 1439. The intensity scale is in linear units, which allows a direct comparison. For GD 806 we can identify hydrogen lines only, and a very weak CH-feature. The sharpness of the lines is in accordance with computations by Wehrse and Liebert (1981) for cool white dwarfs with

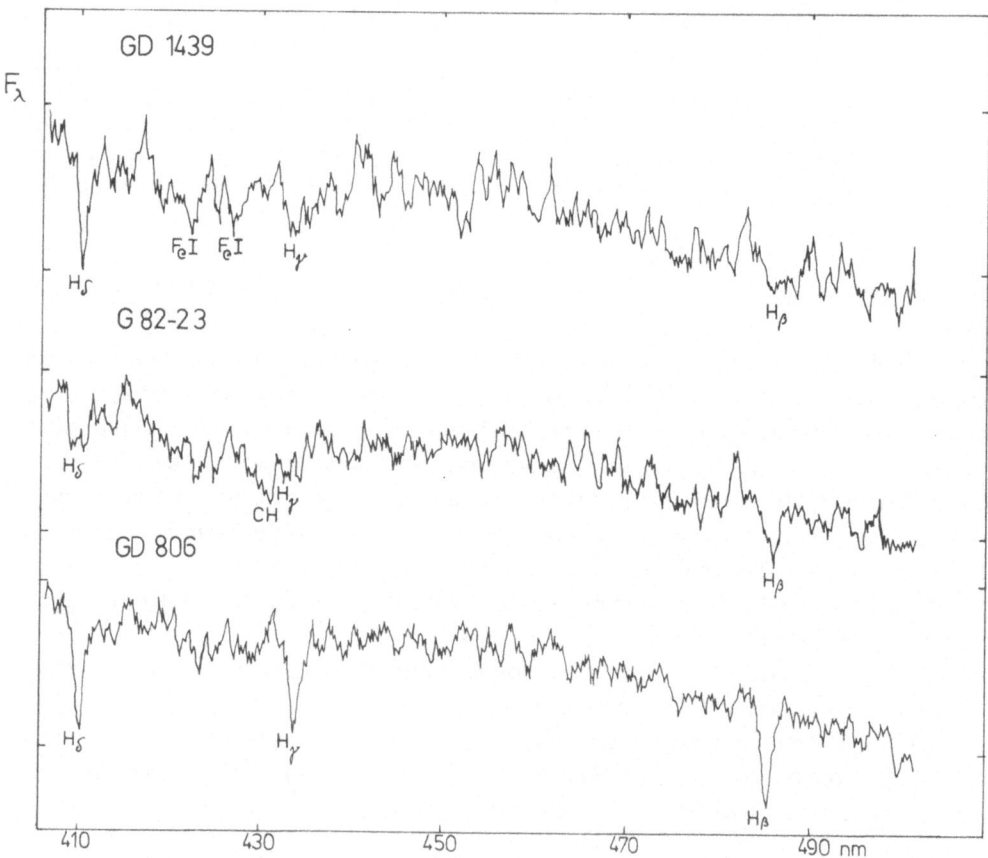

Fig. 2.   Image tube spectrum of G 82–23 compared to a white dwarf (GD 806) and a subdwarf (GD 1439). The intensity scale is in linear units.

atmospheres of nearly pure hydrogen and the calculations by Bessell and Wickrama-singhe (1979).

The spectrum of GD 1439 shows a steeper gradient in the red wavelength range, strong Fe I lines and a weak G band of CH whereas the equivalent widths of the Balmer lines are comparable to those of GD 806. G 82–23 apparently has weaker Balmer lines with $H\gamma$ merging into the strong CH-band, very weak Fe I lines are visible, no other features. The second spectrum differs in the wavelength range 400–430 nm, $H\delta$ is sharper, $H\gamma$ stronger with a slight decrease of the G band. Unfortunately, the radial velocities could not be obtained with high accuracy, due to the shape of the Balmer lines, they varied between 55 km s^{-1} for $H\beta$ to 190 km s^{-1} for $H\delta$ in the other spectrum with no significant systematic difference between the two spectra. More spectra covering a larger time interval would be necessary.

A spectrum most similar to G 82–23 in Figure 2 has been published by Liebert (1976) for GH 7–21 which showed a very strong CH band and additional features of C_2 which are missing in our star. GH 7–21 has been classified as a subluminous object by the author due to the band strength and the visibility of hydrogen lines. A parallax is not known for this star.

3. Conclusions

From our observations – photometry and spectrophotometry – and model atmosphere computations for white dwarfs and subdwarfs it is not possible to present a conclusive fit for G 82–23 for *one* star with the measured trigonometric parallax of $\pi = ''.0033$. A subdwarf with a CH band of the observed equivalent width would have a $B - V$ of 0.8 or even larger in correspondence to the M_V of $7^{m}.4$. The $(U - V)$ and especially the Strömgren colours in connection with $(R - I)$ would suggest a hotter star. A subdwarf of this colour, however, could not show band features due to high temperature and corresponding low pressure in a hydrogen-rich atmosphere. In Figure 3 of the paper by Bessell and Wickramasinghe (1979) G 82–23 would be situated above the locus of line-free subdwarfs.

That is the reason why we think that our observed differences in the Strömgren colours are real and due to a binary effect, where the hotter component is a featureless (DC) white dwarf with an effective temperature $\sim 11\,000$ K like EG 131 ($B - V = 0.07$; $U - B = -0.81$, $M_V \sim V = 12.29$). This star shows very weak He I lines ($\lambda = 389$ nm, 588 nm) in the visible and strong lines ($\lambda 193$ nm, 165 nm) in the far UV (Wegner, 1981). Cooler stars of this composition usually show the Swan-band of C_2 only. The second component of G 82–23 is the main contributor to our observed spectral features in Figure 2 and should be an extremely metal-poor subdwarf like GH 7–21 or G 14–24. With the parallax for the latter as measured by Harrington and Dahn (1980), $\pi = 0''.0126$ and the corresponding $M_v = 8^{m}.32$ we calculated the colours of the system: $B - V = 0.71$, $U - B = -0.10$, $b - y = 0.63$, $u - b = 1.19$.

The agreement is satisfactory for the Johnson colours. For the Strömgren colours, a better fit in colours can be obtained with a hotter helium-rich white dwarf like EG 145 or LP 49–275. With the data by Harrington and collaborators the colours of the system are $B - V = 0.68$, $U - B = -0.18$, $b - y = 0.51$, $u - b = 1.19$ (EG 145), and $B - V = 0.71$, $U - B = -0.17$, $b - y = 0.50$, $u - b = 1.15$ (LP 49–275), respectively. These calculations strongly depend on the accuracy of the parallax for both stars and for G 82–23. The new value by Van Altena (1984) would suggest a more metal-rich subdwarf, which is not seen in all our observations.

Within the accuracy limits of all the parallax determinations, our assumption of a combined system seems to be the most reliable. Additional spectra and more photometry data will be taken within our programme.

Acknowledgements

We thank Dr H. Jahreiss for making available the new parallax for G 82–23 and Dr W. Gliese for discussions.

References

Bessell, M. and Wickramasinghe, D. T.: 1979, *Astrophys. J.* **227**, 232.
Bues, I. and Rupprecht, G.: 1984, *Mitt. Astron. Ges.* **62**, 263.
Eggen, O. J. Greenstein, J. L.: 1965, *Astrophys. J.* **142**, 925.
Giclas, H. L., Burnham, R., and Thomas, N. E.: 1978, *Lowell Observ. Bull.* **164**.
Giclas, H. L., Burnham, R., and Thomas, N. E.: 1980, *Lowell Observ. Bull.* **166**.
Gliese, W.: 1969, *Catalogue of Nearby Stars*, 2nd edition, Veröffentl. Astron. Rechen-Inst., Heidelberg, No. 22.
Greenstein, J. L.: 1976, *Astron. J.* **81**, 323.
Greenstein, J. L.: 1984, *Publ. Astron. Soc. Pacific* **96**, 62.
Harrington, R. S. and Dahn, C. C.: 1980, *Astron. J.* **85**, 454.
Lacombe, P. and Fontaine, G.: 1981, *Astron. Astrophys. Suppl.* **43**, 367.
Liebert, J.: 1976, *Astrophys. J.* **204**, L93.
Liebert, J., Dahn, C. C., Gresham, M., and Strittmatter, P. A.: 1979, *Astrophys. J.* **233**, 226.
Rupprecht, G.: 1983, Ph.D. Thesis, Erlangen.
Rupprecht, G. and Bues, I.: 1983, *ESO Messenger* **34**, 24.
Van Altena, W. F.: 1984, Preprint *General Catalogue of Trigonometric Stellar Parallaxes*, 3rd edition.
Wegner, G.: 1981, *Astrophys. J.* **245**, L27.
Wegner, G.: 1983, *Astron. J.* **88**, 109.
Wehrse, R. and Liebert, J.: 1980, *Astron. Astrophys.* **83**, 184.

ACCURACY OF CLOSE BINARY MASS DETERMINATIONS
FROM PARALLAXES*

W. VAN HAMME** AND R. E. WILSON

Department of Astronomy, University of Florida, Gainesville, U.S.A.

(Received 18 September, 1984)

Abstract. Absolute masses for W Ursae Majoris and Algol-type close binaries can be determined from their parallax, if observed, and the relative sizes of the stars and their mass ratio, obtained from a light curve solution. An error propagation study compares the typical order of magnitude of the various terms involved, and shows how accurate parallaxes have to be in order to make the procedure work, i.e., making the parallax term not larger than the combined non-parallax terms, and producing reasonably low mass errors. Some comments are made on the possibilities with respect to the HIPPARCOS program.

1. Introduction

Observed trigonometric parallaxes for certain types of eclipsing close binaries, when combined with certain purely photometric data and light curve parameters are, in principle, sufficient to determine absolute masses. Intuitively, if we know the distance, we know the luminosity. A suitable temperature indicator (such as $B - V$) then permits calculation of a star's area and thus its linear dimensions. Since the Roche model relates the size of a lobe-filling star to the orbital semi-major axis (a), we find a, which is the final parameter needed in Kepler's third law so as to allow calculation of absolute masses. It is presumed that the mass ratio q, which also is needed in Kepler's third law, has been found by accurate light curve analysis[†]. So in cases where radial velocities are not available or where they give very unreliable results (30 to 50% mass errors are not unusual – viz., e.g., Popper, 1980), observing the parallax could save the situation. The required mass ratios can be found from light curves when we are dealing with:

(1) Semi-detached (Algol) systems, for which the secondary (less massive) component exactly fills its Roche lobe. From the light curve the relative sizes of the components can be found. The size of the secondary is the same as that of its Roche lobe, which completely determines the mass ratio.

(2) Contact binaries (i.e., binaries with both components overfilling their Roche lobes), with both stellar surfaces at the same potential. The figure of the common envelope uniquely determines the mass ratio.

---

* Communication presented at the International Conference on 'Astrometric Binaries', held on 13–15 June, 1984, at the Remeis-Sternwarte Bamberg, Germany, to commemorate the 200th anniversary of the birth of Friedrich Wilhelm Bessel (1784–1846).
** On leave from Rijksuniversiteit Gent, Sterrenkundig Observatorium, Gent, Belgium.

[†] Some of the same relations were used by Huang (1962), Woolf (1965), and Wilson (1974) to estimate the mass ratio of β Lyrae, except that those authors used a spectroscopically measured orbital semi-major axis to derive a mass ratio, while we use a measured mass ratio to derive the semi-major axis (and thus the absolute masses).

Astrophysics and Space Science **110** (1985) 169–175. 0004–640X/85.15
© 1985 *by D. Reidel Publishing Company.*

In the next section we show how the mass, m_1, of the primary component of a close binary system can be calculated from observed quantities (the parallax π, the V-magnitude, and the $B - V$ color index), and parameters resulting from a light curve solution (the relative radius r, mass ratio q, and orbital period P).

In Section 3 we show how the uncertainties in the various parameters propagate. Our aim is to estimate the limits of usefulness of this method for a sample of observed binaries. In particular, we are interested in the parallax accuracy required to give reasonably accurate and useful absolute masses. Some conclusions are given in Section 4.

2. The Method

Using Kepler's third law,

$$G(m_1 + m_2) = (2\pi/P)^2 \, a^3 \,,$$

in which G is the gravitational constant, m_1 and m_2 the masses of the components, a their orbital separation, and P the period, and introducing the mass ratio $q = m_2/m_1$, we can write

$$m_1 = \frac{4\pi^2}{G} \frac{a^3}{P^2(1 + q)} \,. \tag{1}$$

The orbital distance a (in solar radii) can be expressed as a function of the V-magnitude, the bolometric correction BC, the parallax π, the relative radius r (i.e., the radius in units of orbital semi-major axis), and the effective temperature T_e, using the well-known relations

$$a = (R/R_\odot)/r \,, \tag{2}$$

$$L/L_\odot = (T_e/T_{e\odot})^4 \, (R/R_\odot)^2 \,, \tag{3}$$

$$\log L/L_\odot = 0.4(M_{\mathrm{bol}\odot} - M_{\mathrm{bol}}) \,, \tag{4}$$

$$M_{\mathrm{bol}} = M_V + BC \,, \tag{5}$$

and

$$M_V = V + 5 + 5 \log \pi \,; \tag{6}$$

where L is the luminosity of the star, M_{bol} its absolute bolometric magnitude, M_V the absolute V-magnitude, and π the parallax in arc seconds.

Combining Equations (2)–(6) we have

$$\mathrm{Log}\, a = (0.2 M_{\mathrm{bol}\odot} + 2 \log T_{e\odot} - 1) - 0.2(BC + V) - \log \pi - 2 \log T_e - \log r \,. \tag{7}$$

Primary components of Algols and W Ursae Majoris binaries can safely be assumed to be Main-Sequence objects, so that proper Main Sequence (bolometric correction–color index) and (effective temperature–color index) relations give BC and $\log T_e$ as a function of $B - V$. So, combining (1) and (7) we finally obtain m_1 as a function of the parallax and photometric quantities

$$m_1 = m_1[P, q, r, V, \pi, (B - V)]. \tag{8}$$

3. Error Propagation Study

The error propagation formula

$$\sigma_{m_1}^2 = \left(\frac{\partial m_1}{\partial P}\right)^2 \sigma_P^2 + \left(\frac{\partial m_1}{\partial q}\right)^2 \sigma_q^2 + \left(\frac{\partial m_1}{\partial V}\right)^2 \sigma_V^2 +$$

$$+ \left(\frac{\partial m_1}{\partial \pi}\right)^2 \sigma_\pi^2 + \left(\frac{\partial m_1}{\partial r}\right)^2 \sigma_r^2 + \left(\frac{\partial m_1}{\partial (B - V)}\right)^2 \sigma_{(B - V)}^2 +$$

$$+ 2 \left(\frac{\partial m_1}{\partial r}\right)\left(\frac{\partial m_1}{\partial q}\right) \sigma_r \sigma_q \rho_{rq},$$

in which the σ's indicate standard deviations, and ρ_{rq} the correlation coefficient between r and q (r and q result from a light curve solution and are correlated), after some minor algebra, operating on the explicit Equation (8), becomes

$$\sigma_{m_1}^2 / m_1^2 = \frac{4}{P^2} \sigma_P^2 + \frac{1}{(1 + q)^2} \sigma_q^2 + (0.6 \ln 10)^2 \sigma_V^2 + \frac{9}{r^2} \sigma_r^2 +$$

$$+ \frac{9}{\pi^2} \sigma_\pi^2 + 9(\ln 10)^2 \left[0.2 \frac{\partial BC}{\partial (B - V)} + 2 \frac{\partial \log T_e}{\partial (B - V)}\right]^2 \sigma_{(B - V)}^2 +$$

$$+ \frac{6}{r(1 + q)} \sigma_r \sigma_q \rho_{rq}. \tag{9}$$

Eight W Ursae Majoris and one Algol system (listed in Table I) were selected for having at least one published light curve solution, so that some of the required quantities and their standard deviations are available. In most cases the V-magnitude and $B - V$ color index are given in the literature without mentioning any errors, so a typical standard deviation, according to Johnson and Morgan (1953), of $0\overset{m}{.}025$ for the V-magnitude and $0\overset{m}{.}013$ for the $B - V$ color index was assumed.

With the exception of AE Phe and AS Eri, for which light curve solutions were obtained by us, ρ_{rq} was not available. However, since the correlation term in (9) amounts to 5.6×10^{-5} for AE Phe and 4.6×10^{-7} for AS Eri, and is a few orders of magnitude smaller than, for instance, the V-magnitude term or the $B - V$ term, we assumed it to be negligible in all other cases.

The derivatives in the $B - V$-term of Equation (9) were calculated according to a

W. VAN HAMME AND R. E. WILSON

TABLE I

Estimated parallaxes and parallax uncertainties

System	σ_π/π if $\sigma_{m_1}/m_1 = 0.15$	π (")	d (pc)	σ_π if $\sigma_{m_1}/m_1 = 0.15$
44i Boo	0.043	0.067[a]	13	0.0034
AE Phe	0.044	0.021	48	0.0009
AW UMa	0.044	0.0092	108	0.0004
V535 Ara	0.044	0.0086	116	0.0004
RZ Com	0.039	0.0081	123	0.0003
TX Cnc	0.043	0.0077	130	0.0003
RZ Tau	0.044	0.0037	270	0.0002
XY Boo	0.039	0.0032	308	0.0001
AS Eri	0.047	0.0056	178	0.0003

[a] Observed trigonometric parallax.

polynomial least squares fit of the $[BC - (B - V)]$ and $[\log T_e - (B - V)]$ Main-Sequence relation of Morton and Adams (1968) and Harris $et\ al.$ (1963). This fit gives

$$\frac{\partial (\log T_e)}{\partial (B - V)} = -118\,350(B - V)^5 - 95\,480(B - V)^4 - 27\,408(B - V)^3 -$$

$$- 3390(B - V)^2 - 170(B - V) - 3.941\,,$$

$$\frac{\partial (BC)}{\partial (B - V)} = 844\,776(B - V)^5 + 713\,515(B - V)^4 + 220\,500(B - V)^3 +$$

$$+ 30\,837(B - V)^2 + 1920(B - V) + 48.7\,,$$

for $-0.32 \le B - V < 0.00$, and

$$\frac{\partial (\log T_e)}{\partial (B - V)} = 17.82(B - V)^5 - 52.05(B - V)^4 + 54.4(B - V)^3 -$$

$$- 24.36(B - V)^2 + 4.38(B - V) - 0.561\,,$$

$$\frac{\partial (BC)}{\partial (B - V)} = -25.5(B - V)^5 + 66.5(B - V)^4 - 66.8(B - V)^3 +$$

$$+ 33.066(B - V)^2 - 10.12(B - V) + 1.45\,,$$

for $0.00 \le B - V \le 1.18$.

Expression (9) can be used to calculate the maximum relative parallax error σ_π/π which gives masses to a prescribed accuracy, say for instance 15%. However, it would be interesting also to have absolute parallax errors, for which an explicit value of the parallax is needed. Since we want to estimate the uncertainties of the procedure, we must establish the order of the distances with which we are dealing. Among the binaries of Table I, only 44i Boo has an observed trigonometric parallax, but with an estimate of the absolute masses and radii of our system's primary components from a radial velocity solution, one can reverse the procedure of Section 2 and estimate parallaxes from masses. This has been done, and the parallaxes and distances obtained this way are also listed in Table I. We emphasize that, except for 44i Boo, these parallaxes are not trigonometric. The distances are intended only as approximations for use in test calculations. Figure 1 shows, for each system, the relative mass standard deviation vs the parallax standard deviation, according to Equation (9), and is based on the parallaxes listed in Table I. Distances are given in parentheses.

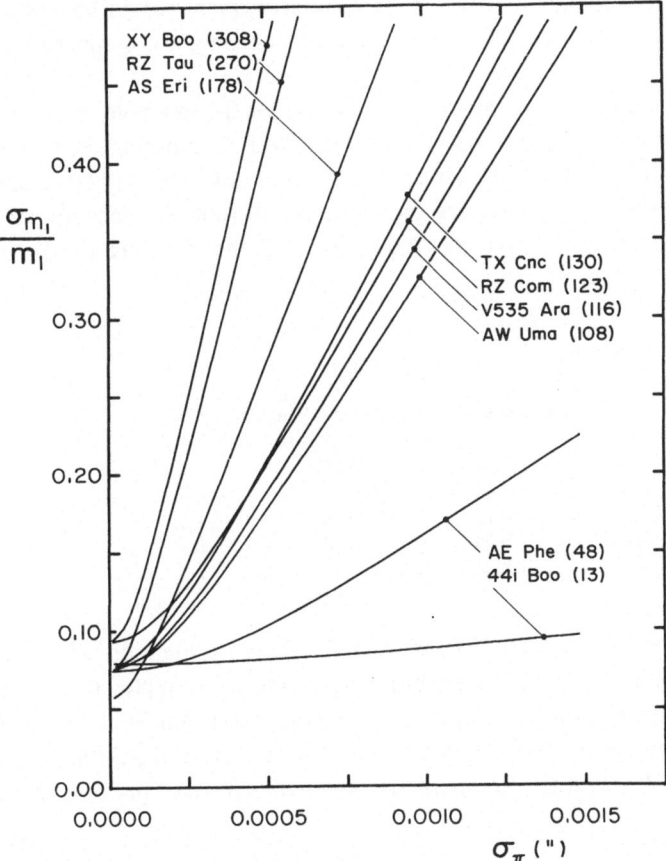

Fig. 1. The relative mass standard deviation vs the parallax standard deviation for selected binaries. Numbers in parentheses indicate approximate distances in pc.

One can see that, due to the typical errors in photometric data (color indices, V-magnitudes), and typical errors of parameters derived from light curves (q, r), one cannot expect mass accuracies better than 5 to 10%, even with virtually an exact parallax. The rate of decrease of mass accuracy with π-error, as reflected by the steepness of the lines in Figure 1, becomes greater with increasing distance. So one can conclude that the method outlined in Section 2 will produce masses with an accuracy better than, say, 15% only for systems within about 100 pc. However, with the predicted parallax accuracy of the HIPPARCOS project ($\sigma_\pi \approx 0\rlap{.}''002$), which may eventually improve by a factor 2 to 4, an interesting region in the diagram of Figure 1, corresponding to mass errors of 20 to 30%, becomes accessible. In traditional mass determinations for close binaries (double-lined spectroscopic and eclipsing systems), mass accuracies of 5 to 10% are only obtained in a few cases. For Algols and contact binaries we may encounter mass errors of 30 to 50%, as already mentioned in Section 1. So the method of Section 2 can serve as a very useful tool for obtaining masses for close binaries which are independent of radial velocities. Further improvement would be realized with more accurate V-magnitudes and $B - V$ color indices, causing a downward shift of all lines in Figure 1.

As a further illustration of this point we consider the example of the very low-mass ratio ($q \approx 0.07$) W Ursae Majoris system AW UMa. Combining the results of the light curve solution of Wilson and Devinney (1973), the radial velocity solutions of McLean (1981), and assuming an effective temperature of 7000 K according to the F0–F2 spectral type given by Paczynski (1964), we find the absolute dimensions given in Table II.

TABLE II

Absolute parameters for AW UMa

	$m/m_\odot$	$R/R_\odot$	$\log L/L_\odot$
Primary	4.16	2.50	1.1
Secondary	0.29	0.84	0.1

These numbers place AW UMa in an H–R-diagram location which is completely incompatible with any standard evolutionary models for single stars. Neither a post-Main Sequence (see, for instance, Iben, 1967), nor a pre-Main-Sequence stage (see Iben, 1965) can explain these numbers. Hence, a very accurate parallax (the system AW UMa is only about 100 pc away), could shed some light on these peculiar results.

Acknowledgements

W.V.H. gratefully acknowledges a travel grant from the National Science Foundation administered by the American Astronomical Society. R.E.W. is pleased to acknowledge support under N.S.F. grant AST 8203700.

References

Harris, D. L., Strand, K. Aa., and Worley, C. E.: 1963, in K. Aa. Strand, *Basic Astronomical Data*, University of Chicago Press, p. 273.
Huang, S. S.: 1962, *Astrophys. J.* **136**, 903.

Iben, I.: 1965, *Astrophys. J.* **141**, 993.
Iben, I.: 1967, *Ann. Rev. Astron. Astrophys.* **5**, 571.
Johnson, H. L. and Morgan, W. W.: 1953, *Astrophys. J.* **117**, 313.
McLean, B. J.: 1981, *Monthly Notices Roy. Astron. Soc.* **195**, 931.
Morton, D. C. and Adams, T. F.: 1968, *Astrophys. J.* **151**, 611.
Paczynski, B.: 1964, *Astron. J.* **69**, 124.
Popper, D. M.: 1980, *Ann. Rev. Astron. Astrophys.* **18**, 115.
Wilson, R. E.: 1974, *Astrophys. J.* **189**, 319.
Wilson, R. E. and Devinney, E. J.: 1973, *Astrophys. J.* **182**, 539.
Woolf, N. J.: 1965, *Astrophys. J.* **141**, 155.

EPSILON AURIGAE*

NASA/Goddard Space Flight Center, Greenbelt, Maryland, U.S.A.

(Received 22 August, 1984)

Abstract. In April 1984, fourth contact ended the two year long eclipse of Epsilon Aurigae. An astrometric study of the study of the system was carried out by Van de Kamp (1978) leading to the conclusion that the orbit is seen very close to edge on. The eclipse was monitored by a number of groups from the ground and from spacecraft such as the IUE. Ultraviolet observations of the system from IUE have thrown new light on the nature of the system that lead us to conclude that the secondary object is probably a cold, dusty accretion disk surrounding a star that is completely hidden inside the disk.

Epsilon Aurigae is a particularly appropriate system to be discussed here both because this conference proceedings is in honor of F. W. Bessel, and because the conference was held in Germany. This unique binary system has been observed for many years both astrometrically and spectroscopically. The light variability of the system was first noted in 1821 by a German amateur astronomer and extensive visual brightness estimates were made at the 1848 eclipse in Bonn by Bessel's student Argelander. Early spectroscopic observations were made at Potsdam and at the Yerkes Observatory. Ludendorff recognized the periodic nature of the system and concluded that it is a spectroscopic binary. In 1924 Ludendorff (1924) published the first spectroscopic orbit of the system. His calculations have been updated and then farther refined by Morris (cf. Morris, 1962; Wright, 1970). The relevant quantities are $P = 9890$ days, $e = 0.200 \pm 0.034$, $a_1 \sin i = 12.9$ AU, and mass function = $3.12 \odot$.

The peculiar nature of the system became clear quite early. No secondary eclipse has ever been observed, suggesting that the secondary is significantly fainter than the primary. During the primary eclipse the light level of the system is reduced by about 0.8 mag. or by about 50% suggesting that either the eclipse is partial or the secondary is smaller than the primary. In fact, both of these possibilities appear to be ruled out by additional observations. The eclipse light curve has the flat bottomed shape typical of a central eclipse and, in addition, the eclipse is quite long; the time period from first contact to fourth contact is roughly 714 days (Figure 1) and totality lasts nearly a year. The length of the eclipse combined with the orbital parameters permits us to calculate the dimensions of the secondary, which must be substantially larger than the primary. Therefore, the primary ought to be totally eclipsed by the larger fainter secondary object, contradicting the depth-of-eclipse observations. There is one additional very important observational fact; during eclipse the spectrum is dimmed uniformly across the entire

* Communication presented at the International Conference on 'Astrometric Binaries', held on 13–15 June, 1984, at the Remeis–Sternwarte Bamberg, Germany, to commemorate the 200th anniversary of the birth of Friedrich Wilhelm Bessel (1784–1846).

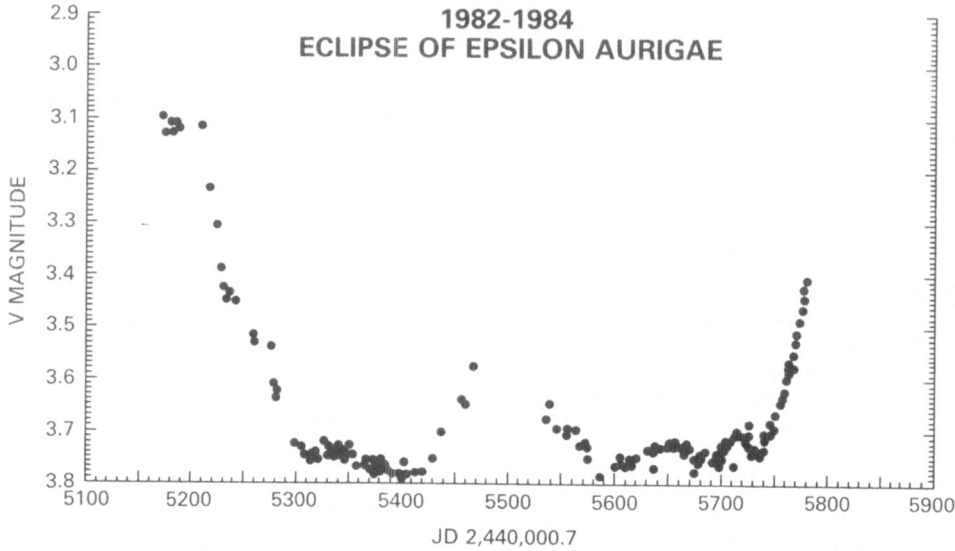

Fig. 1.   The visible light curve of ε Aurigae during the recent eclipse. The numerical values of the magnitudes are taken from various observers as reported in the ε Aurigae *Newsletter* and should be viewed as provisional.

visible light observations are summarized in a number of reviews, including Wright (1970) and Sahade and Wood (1978).

There have been several models proposed to explain these unusual observations (see, for instance, the review of Sahade and Wood, 1978). Kuiper *et al.* (1937) asserted that the secondary is a very large, cool infrared I-star that is partially transparent. In that model, the gray absorption would have been caused by electron scattering in a layer of the I-star that is ionized by the F-supergiant. It is difficult to understand how one could avoid additional non-gray continuous absorptions in such a system; if the material is ionized by phototons from the primary, then why do we not see photoionization continua in the spectrum? New 20-micron infrared observations of the system (Backman *et al.*, 1984) do show the presence of a large object with a temperature of about 500 K. A way out of the non-gray absorption problems posed by a semi-transparent secondary is to accept Huang's (1965) model of a flat, rotating disk that is opaque when viewed edge on. The extent of the disk in the orbital plane is sufficient to explain the length of the eclipse (about 1000 solar radii), but the extent perpendicular to the orbital plane is less than the diameter of the supergiant so that, even at mid eclipse, the 'poles' of the primary are observed 'above' and 'below' the disk. The projected area of the disk is consistent with the IR observations (Backman *et al.*, 1984). If we accept the existence of a disk, then we must try to ascertain its nature. Kopal (1954, 1971, 1972) suggested that the secondary disk is a semi-transparent ring of solid particles surrounding a secondary component. Handbury and Williams (1976) have expanded on this idea and speculated that the system might be very young, with the primary being a pre-Main-Sequence star and the secondary being a embryonic solar system. If these were true, then one should see measurable changes in the massive primary on time-scales of decades, based on the predicted evolutionary rates of such objects. Parsons (1984) has made a careful study of high-dispersion spectra from the 1890's used by Maury and Pickering (1897) and found no convincing evidence of spectral changes. These observations appear to rule out the primary as a pre-Main-Sequence star. We will briefly introduce an alternative model below.

A detailed study of Epsilon Aurigae as an astrometric binary has been carried out by Van de Kamp (1978), who used plate material covering the years 1939 to 1977. He found the angular elements

$$i = 89° \pm 3° , \qquad \Omega = 92° \pm 3°$$

and

$$a = 0\rlap{.}''0227 \pm 0\rlap{.}''0010 .$$

Van de Kamp's inclination is significantly different from an earlier value of 72° found by Strand (1959), who used plate material covering the period 1962 to 1958. This latter value for i would lead to a grazing eclipse. As far as the physical dimensions of the system are concerned, there is little difference between 72° and 89° since the relevant quantities are $\sin i$ and $\sin^3 i$, which differ by less than the inclinations themselves. Strand's (1959) analysis is published only as an abstract and was never followed up by a referred paper; so that it is difficult to evaluate his results. The linear semi-major axis from the spectroscopic solution of the system and the angular semi-major axis from the astrometric solution can be combined to yield an absolute parallax

$$\pi = 0\rlap{.}''001\,72 \pm 0\rlap{.}''000\,08 ,$$

which leads to an absolute visual magnitude $M = -6.7$ corrected for interstellar absorption (Van de Kamp, 1978). Schmidt-Kaler (1961) has made a careful study of these rare late-type supergiants in galactic clusters and associations and concludes that the absolute magnitude of an F0 Ia star is -8.5. Bouw and Parsons (1971) have extended this work, based on a somewhat larger set of observational data and arrive at a similar value, though their work shows that the inferred absolute magnitude values exhibit a total scatter of slightly over a magnitude. Given all the uncertainties in both the canonical M for F0 Ia and the value for Epsilon Aurigae, we conclude that the values are not inconsistent. Kuiper et al. (1937) favoured an inclination close to 70°. Their argument was based on the absolute magnitudes that were inferred from various orbit solutions. As it turns out, the greater baseline used by Morris and by Van de Kamp has led to better orbital solution. Together with the more recent determinations of the canonical values for supergiants the absolute magnitudes no longer presents such a difficulty, even with an assumed inclination of 90°. Therefore, we will assume Van de Kamp's value here.

The most recent eclipse of Epsilon Aurigae began with first contact in July, 1982 and ended with fourth contact in May, 1984. Observations with the International Ultraviolet Explorer (IUE) have revealed at least two very interesting results (Altner et al., 1984). The first result was found by intercomparing the shapes of the low-dispersion spectra

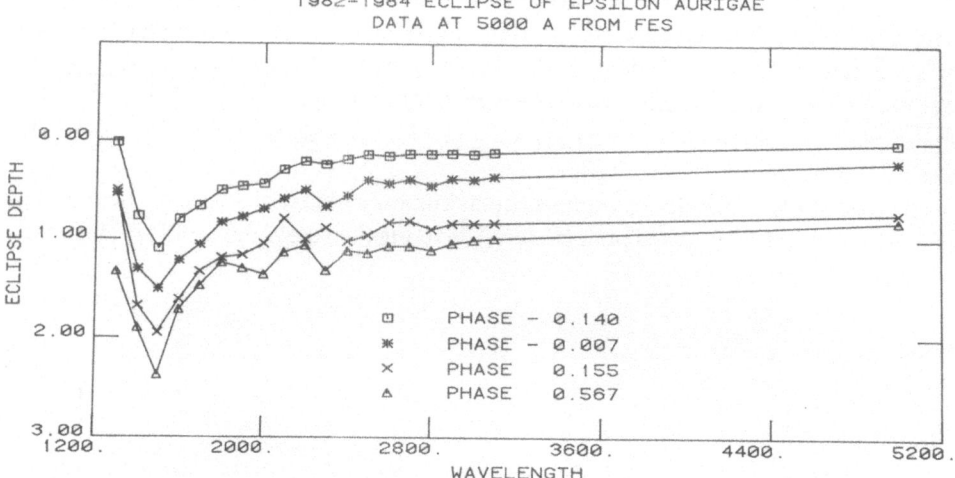

Fig. 2.   The 1982–1984 eclipse of ε Aurigae as observed by IUE. The ordinate is depth of eclipse using a
spectrum obtained on 4 April, 1982 by T. Ake and T. Simon as the out-of-eclipse fiducial. The UV data is
from low dispersion spectra and was averaged into 100 Å bins. The 5000 Å point is from the IUE Fine Error
Sensor. Note that the eclipse is essentially gray longward of about 2400 Å.

taken at various phases during the eclipse. Figure 2 shows a plot of eclipse depth vs
wavelength for several typical phases covering ingress and totality. The pre-eclipse
reference spectrum was obtained by T. Ake and T. Simon on 4 April, 1982. Note that
the depth is nearly wavelength independent at all phases from about 2400 Å longward
to 5000 Å. This observation is consistent with the earlier results from ground-based
studies. Between 2400 and 1500 Å, the eclipse depth increases with decreasing wavelength. Chapman *et al.* (1983) have suggested that the extra source of opacity at short
wavelengths may be due to dust in the vicinity of the secondary object, particularly in
the outer regions where the material density is low and the disk may not be fully opaque.
In their paper, Chapman *et al.* (1983) used a spectrum obtained on 13 April, 1982 as
the out-of-eclipse reference and a spectrum obtained on 21 September, 1982 as the
in-eclipse spectrum. Two papers (Boehm *et al.*, 1984; and Parthasarathy and Lambert,
1983) take issue with the dust interpretation. Those authors assert that the 13 April,
1983 spectrum was obtained at a time when the supergiant star was 'active' and the
21 September, 1982 spectrum was taken at a quiescent period, then the apparent
deepening of the eclipse is dominated by intrinsic variability of the supergiant. The
4 April, 1982 spectrum used as a fiducial in the newer analysis shown in Figure 2 was
not obtained at an 'active' period and the Chapman *et al.* (1983) result remains valid.
The decrease in eclipse depth shortward of 1500 Å may be due to the fact that the shorter
wavelength emission is dominated by a hot object imbedded in the secondary (Boehm
et al., 1984). The second result from the IUE observations concerns the Mg II resonance
lines which show a very broad absorption feature, superposed on which appear to be
lines with P Cygni profiles at the expected wavelengths of the two members of the

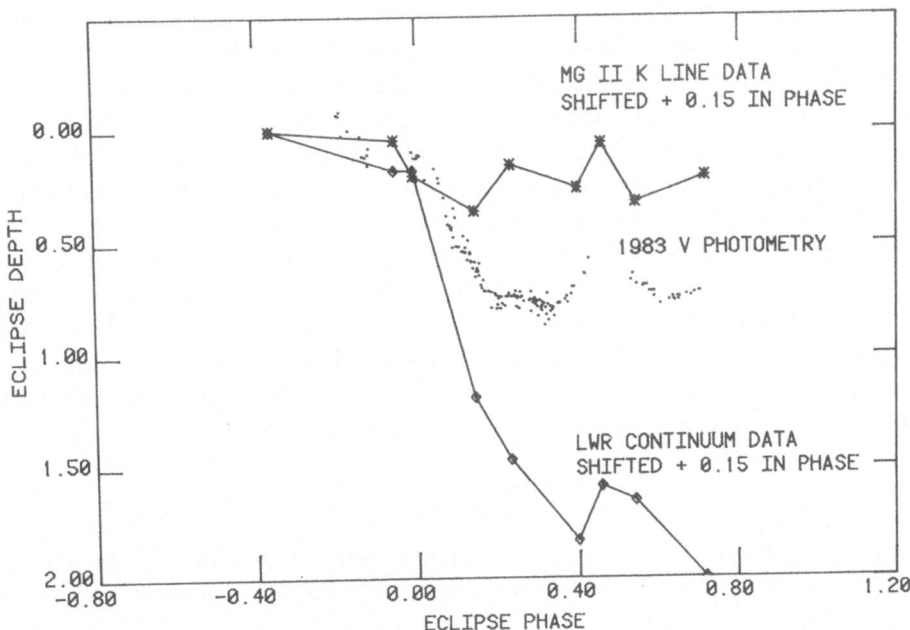

Fig. 3. The change in flux of the MgII K line as a function of eclipse phase compared with a point in the nearby continuum. Note that the flux in the MgII emission is affected very little by the eclipse. The V-magnitudes are plotted on the same graph, but have been shifted left by 100 days. The UV eclipse, as illustrated by the continuum data is, therefore, seen to be about 100 days earlier than the visible eclipse.

doublet. In fact, we believe that we are not seeing P Cygni lines but rather emission which arises somewhere in the stellar system with a narrow, blueward interstellar absorption. Casual inspection of the MgII resonance lines shows an increase in the relative intensity of the emission as the eclipse progresses. A more careful study shows the result in Figure 3. The asterisks are the intensity of the MgII K line as a function of phase while the diamonds are the intensity in the nearby continuum. The eclipse phase is calculated such that phase 0.0 is at the time of first contact and phase 1.0 at the time of fourth contact, using Gyldenkerne's 1970 prediction times of the contact. Plotted on the same graph is the V magnitude from the 1983 eclipse. Note that one must shift the UV data by +0.15 in phase to make the steepest declines in visible and UV light correspond; that is, the UV eclipse starts 0.15 in phase earlier than the visible light eclipse. The fact that MgII emission does not change significantly in intensity means that the source of the emission is not eclipsed by the secondary. Parthasarathy and Lambert (1983) argue that the MgII emission comes from the chromosphere of the supergiant and arises mainly from the limb. If that was the case, one can visualize a situation where the fraction of the limb that is eclipsed is less than the fraction of the disk that is eclipsed.

Using Morris' mass function and his estimated mass of 15.5 solar masses for the primary, the mass of the secondary is roughly 13.7 solar masses. I believe that this secondary may be a fossil accretion disk around a very compact star. The disk was formed at an earlier stage in the evolution of the system when the primary was losing mass at a high rate and after the secondary had evolved into a condensed object, perhaps a black hole. Today, the rate of accretion onto the secondary is so slow that the energetic phenomena usually though to be characteristic of such systems are not observed. The disk has cooled to the observed 500 K and small dust grains have condensed. Following the analysis of McCluskey and Kondo (1971), we have addressed ourselves to the question: could the present secondary have been much more massive in the past, thereby evolving into a compact object and losing significant mass in the process and ending with the present masses? If the present secondary could have started as say an 18 solar mass star, and if it could evolve into a compact object without too violent a supernova explosion, then the answer to the question is yes. A more detailed analysis of the situation is under way now and will be published in the near future.

References

Altner, B. M., Chapman, R. D., Kondo, Y., and Stencel, R. E.: 1984, in J. M. Mead, R. D. Chapman, and Y. Kondo (eds.), *The Future of Ultraviolet Astronomy Based on Six Years of IUE Research*, NASA CP (in press).

Backman, D. E., Becklin, E. E., Cruikshank, D. P., Joyce, R. R., Simon, T., and Tokunaga, A.: 1984, preprint.

Boehm, C., Ferluga, S., and Hack, M.: 1984, *Astron. Astrophys.* **130**, 419.

Bouw, G. D. and Parsons, S. B.: 1971, in M. Hack (ed.), *Colloquium on Supergiant Stars*, Proc. 3rd Colloq. on Astrophysics, Trieste, p. 22.

Chapman, R. D., Kondo, Y., and Stencel, R. E.: 1983, *Astrophys. J.* **269**, L17.

Gyldenkerne, K.: 1970, *Vistas Astron.* **12**, 199.

Handbury, M. J. and Williams, I. P.: 1976, *Astrophys. Space Sci.* **45**, 439.

Huang, S.-S.: 1965, *Astrophys. J.* **141**, 976.

Kopal, Z.: 1954, *Observatory* **74**, 14.

Kopal, Z.: 1971, *Astrophys. Space Sci.* **10**, 332.

Kopal, Z.: 1972, in *Nobel Symp. Stockholm*, No. 21, pp. 39–47.

Kuiper, G. P., Struve, O., and Strömgren, B.: 1937, *Astrophys. J.* **86**, 570.

Ludendorff, H.: 1924, *Sitzber. Berlin Preuss. Akad. Wiss.* **9**, 49.

McCluskey, G. E. and Kondo, Y.: 1971, *Astrophys. Space Sci.* **10**, 464.

Maury, A. C. and Pickering, E. C.: 1897, *Harvard Ann.* **28**.

Morris, S. C.: 1962, *J. Roy. Astron. Soc. Canada* **56**, 210.

Parsons, S. B.: 1984, private communication.

Parthasarathy, M. and Lambert, D. L.: 1983, *Publ. Astron. Soc. Pacific* **95**, 1012.

Sahade, J. and Wood, F. B.: 1978, *Interacting Binary Stars*, Pergamon Press, New York, p. 152.

Schmidt-Kaler, T.: 1961, *Z. Astrophys.* **53**, 1, 28.

Strand, K. Aa.: 1959, *Astron. J.* **64**, 346.

Van de Kamp, P.: 1978, *Astron. J.* **83**, 975.

Wright, K. O.: 1970, *Vistas Astron.* **12**, 147.

THE HELIOMETER PRINCIPLE AND SOME MODERN APPLICATIONS*

EDWARD H. GEYER

Sternwarte der Universität Bonn, Observatorium Hoher List, Daun, F.R.G.

(Received 20 July, 1984)

"Astronomy gains only by new results if these are unambiguously obtained. Not the premature guessing but the fundamental acquisition of data and knowledge must be the topic of the efforts."

F. W. Bessel: 1844, *Astron Nachr.* **22**, 145.

Abstract. Beside some historical notes about the large Fraunhofer's heliometer used by Bessel and Argelander, some modern applications of the heliometer principle for the geometric and photometric autocalibration of detectors and the determination of absolute radial velocities with slitless field spectrographs are presented.

1. Introduction

It is sometimes the fate of fundamental inventions made by our ancestors to appear redundant when technology and knowledge has progressed, and are re-invented after quite a while to improve or make the new technique more accurate. This seems to be the case with the 'heliometer principle', an autocalibration technique based on the double-image method for the measurement of angular distances and diameters of celestial bodies.

Out of the need to determine the diameters in any arbitrary direction the 'heliometer' was invented in the middle of the 18th century and perfected by Fraunhofer. In the hands of Bessel and other astronomers of the 19th century like E. Hartwig, most prominent and accurate results for astrometric measurements by visual methods were achieved. The reason for which is that it is based on an autocorrelation-null method. Though photographic astrometry apparently has made it obsolete, its basic principle can be nowadays applied for the geometrical and photometrical autocalibration of panoramic detectors, especially the photographic plate or film, and for the determination of absolute radial velocities with slitless field spectrographs.

2. The Heliometer Principle and its History

Image doubling in a heliometer is performed by using two imaginary optical systems of identical focal length, aperture and form of their entrance pupils, and which are ideally

---

* Communication presented at the International Conference on 'Astrometric Binaries', held on 13–15 June, 1984, at the Remeis-Sternwarte Bamberg, Germany, to commemorate the 200th anniversary of the birth of Friedrich Wilhelm Bessel (1784–1846).

mounted in a way that their optical centers are on the surface of a cylinder, the radius of which is their common focal length, their optical axis being its radii. In the case the optic apertures are circular, their minimum lateral distance d_m of their centers is of course equal to the diameter A of their apertures (Figure 1). Under these conditions, the structure of the Airy's discs of so formed double images of a point source are identical, having the same focal ratio. Seen from the center of symmetry of such an optical arrangement the double images spans an angle 2ϑ, which is given by

$$\tan 2\vartheta = \frac{d}{f} \leq \frac{A}{f} \; , \tag{1}$$

where f is the focal length of the lenses. For $d_m \approx A$ the minimum angle d_m is given by the f-number of the objective lenses. If the latter is kept constant, ϑ_m is independent of A. Therefore, such an optical double image device must not necessarily be placed at the entrance pupil of the telescope but as well can be transferred to its back end if the light beam is re-collimated either before or after the telescopes focal plane (a so-called 'a-focal' system). In the latter case such a design has the advantage that a field lens can be introduced by which the whole useful field of the telescope is made accessible. Such an arrangement, which resembles a Kellner eyepiece, is called a 'focal reducer' (Figure 2).

Its advantages are threefold:

(a) The telescope focal plane can be made accessible for the application of the multi-diaphragm masking technique for sky background and overlapping reduction and for the observation of faint objects in the vicinity of bright ones (Geyer and Schmidt, 1976; Geyer *et al.*, 1979).

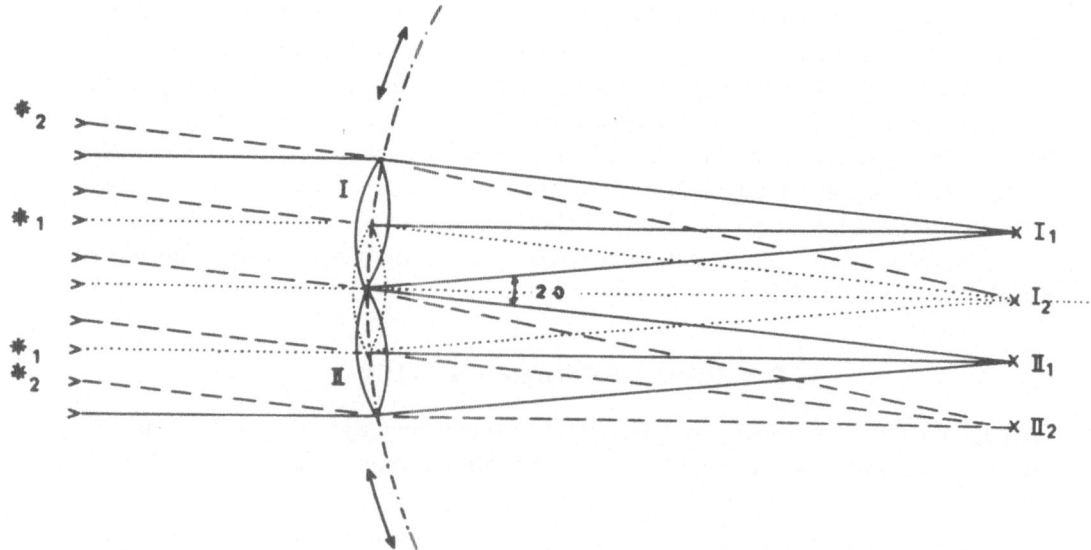

Fig. 1.   The basic layout of a classical heliometer.

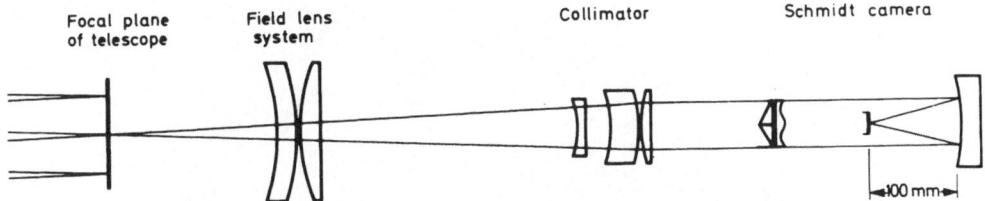

Fig. 2. The principal optical arrangement of a focal reducer system.

(b) The distance between the last optical surface of the collimator and the exit pupil of the whole system can conveniently be made large enough for the installation of dispersing elements (e.g., prisms) and filters with diameters not exceeding 80 mm, as the exil pupil is of this size or smaller.

(c) The focal reduction factor, which also reduces the linear diameter of the seeing disc of the final image, is given by the ratio of the f-number of the telescope to the f-number of the camera, thus making possible the application of modern but small sized, panoramic detectors for fields of view up to one degree also with large telescopes.

If d is kept fixed, the advantage of such a double image device is that for each object within the telescope field the mid position of relevant double images are determined with higher accuracy than the single images. Yet, what is even more important, an auto-calibration of the geometrical distortion in the none optimal matching of the detector to the focal surface is achieved. In other words, local or systematic de-focusing effects by the surface irregularities of the panoramic detectors can be fully taken into account. This is especially important for astrometric work with photographic plates as the variation of the emulsion thickness and the topologic effects by its processing cannot be determined otherwise.

The first double-image micrometer with a fixed minimum distance for the optical centers was designed by Bouguer (1748) for the determination of the annual variation of the diameter of the solar disc. He named his instrument 'héliomètre' (Figure 3). The basic idea seems to go back to Ole Römer (1675).

To avoid the condition for a minimum angular distance 2ϑ, Dollond (1753) proposed to use only one objective lens, but have it sliced diametrically into exact halfs, making the semi-lenses symmetrically movable along the cutting line and the whole objective head turnable around the optical axis for the alignment of two objects in position angle with their relevant double images (Figure 4). In this way, two objects within the field of view of the telescope can be brought into coincidence of the mutual double images or in another convenient measuring position (e.g., limb contact of the double image of a planetary disc). The angular distance is then given by the measured d, multiplied by the scale factor, which is according to Equation (1) the reciprocal of the effective focal length.

Besides the advantage of the larger angular distance which can be covered by this type of double-image micrometer (up to $1°5$, given by the useful field of the semi-lenses), it offers additional ones, about which Bessel (1812) already was fully aware after having

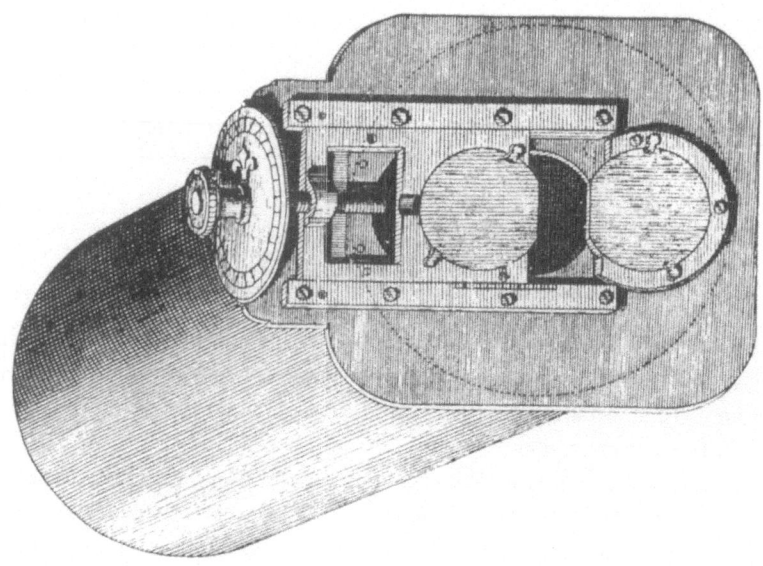

Fig. 3.   Bouguer's 'minimum distance' heliometer (1748).

determined the positions of the great comet of 1811 and of the binary 61 Cyg by means of an improved Dollond heliometer with an accuracy superior to all other astronomical relative position measurements of that time (see also Bessel, 1831):

(a) No field- or/and crosswire illumination is necessary.

(b) The accuracy of the telescope drive has no influence upon the accuracy of the relative position measurements of two objects within the field of view, as their double images are affected in the same way.

(c) Most important, the seeing is influencing the double images also in the same way and is cancelling completely in the case of image coincidence as long as the angular

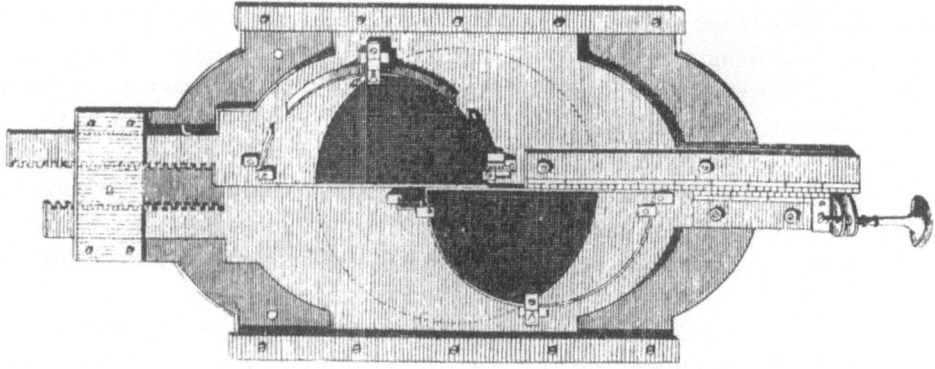

Fig. 4.   Dollond's heliometer head (1755).

distance of two objects is within the isoplanatic field (as we know nowadays). Under these circumstances the seeing pattern of the two sources is identical and, therefore, is of no influence on the image coincidence either.

If a constant angular distance ϑ for the double images is desired and/or sufficient, as is the case for the autocalibration of panoramic detectors, an independent method for producing double images exist by means of a bi-prism (or optical double wedge) with very small, but identical, refracting angles, placed at the entrance pupil of the camera, dividing it again into exact halfs (Figure 2). This avoids the slicing of the camera optics. Such a bi-wedge can be produced up to sizes of 100 cm, so that they can also be used in connection with larger Schmidt cameras.

The double images produced by a bi-prism with small refracting angle $\phi \ll 1$ and refracting index n have an angular separation 2ε under the minimum deviation condition given by

$$2\varepsilon \simeq (n - 1)\phi \text{ radians}. \tag{2}$$

It is independently of the camera geometry if positioned at its entrance pupil. Thus, it can be favourably used for the focal length calibration and/or the geometry of the detector, as already outlined, since the linear distance 2Δ of the double images at the camera's focal surface is given by

$$2\Delta = 2\varepsilon f. \tag{3}$$

The minimum condition can be only fulfilled for the optical axis. Therefore, a slight distortion error for the angular distance of the double images is introduced over the field of the camera, which can be accounted for (see, e.g., Nelles and Geyer, 1981).

Furthermore, as the refracting index $n = n(\lambda, T)$ is a function of the wavelength λ and temperature T, also $\varepsilon = \varepsilon(\lambda, T)$: i.e.,

$$\frac{\partial \varepsilon}{\partial \lambda} = \frac{\partial n}{\partial T} \phi. \tag{4}$$

Both derivatives $\partial n/\partial \lambda$, $\partial n/\partial T$ are well known from laboratory measurements and their influence on 2ε can be accounted for. Of course, over larger wavelength regions the double images are small, identical, spectra of *opposite* direction of dispersion. For the double image-mid position, therefore, is not influenced at all by λ and T, and only by the very small field distortion. In Table I are listed some numerical values for a BK7-bi-prism ($n \simeq 1.52$), $\Delta\lambda = 100$ nm and the Hoher List Observatory Schmidt camera ($f = 1370$ mm).

Of course, by dividing or slicing the entrance or exit pupil of an optical system introduces some drawbacks: the diffraction pattern (Airy disc) of the half-pupil is no longer circularly symmetric, its effective f-number is increased by $\sqrt{2}$, and also for point sources the detection limit is shifted by 0.75 mag. to brighter ones. Finally, due to mechanical reasons the wave front of a point source after passing the two half-lenses is no longer in phase, even if their optical centers are coinciding.

E. H. GEYER

TABLE I

Prismatic image extension $d\Delta/d\lambda$ and linear double image distances Δ of a BK7 bi-wedge and camera focal length $f = 1370$ mm

ϕ (arc sec)	$\lambda = 500$ nm		$\lambda = 400$ nm	
	2Δ (mm)	$d\Delta/d\lambda$ (mm/100 nm)	2Δ (mm)	$d\Delta/d\lambda$ (mm/100 nm)
15	0.100	0.006$_5$	0.106	0.013
30	0.200	0.013	0.212	0.027

The historical development of the 'classical heliometer' as a double-image micrometer and also the observational techniques can be found in the articles by Ambronn (1899), Repsold (1908), and Schur (1898). A paper about the origin of the heliometer was recently presented by Fanque (1983). Therefore, I will restrict to some additional historical remarks concerning to the heliometers by Fraunhofer and used by Bessel.

The Königsberg observatory was built in the years 1809 to 1812–1913. Napoleon, who visited in 1812 the nearly completed building, exclaimed after its purpose was explained to him: "By God, has the king of Prussia still time to think about such things"?

In 1812, Bessel observed for the first time the visual binary 61 Cyg with an improved Dollond heliometer for the determination of the relative position of the components in comparison to neighbouring stars, and developed the concept of the 'astrometric binaries': constant proper motion of the center of mass of the double star; variable motion of the components due to the annual parallactic motion and their absolute orbital motion leading to the determination of the individual masses of the components.

In 1815 Fraunhofer designs in a perfect manner a series of smaller heliometers ($A = 76$ mm, $f = 1150$ mm) for the observatories Berlin, Breslau, Göttingen, and Gotha, and which were the prototypes for all later designs. In 1874 and 1882 they were still extensively used for the Venus passage expeditions.

At the end of 1824 Bessel visited Fraunhofer at Munich and ordered the large Königsberg heliometer ($A = 158$ mm, $f = 2560$ mm). He proposed to him to have the half-lenses moved on the cylinder surface, but Fraunhofer declined on account of mechanical difficulties and promised to have the lenses specially corrected to reduce the out-of-axis distortion.

Shortly before the premature death of Fraunhofer on 7 June, 1826 he had completed the lens for the Königsberg instrument and for safety's sake two additional objectives of nearly identical size, but he could not start with the slicing. This was carried out in 1827 by his collaborators Merz and Mahler successfully for all 3 lenses. The two additional objectives were later on used for the Bonn (1840) and Pulkovo (1841) heliometers. The Königsberg and Pulkovo instruments seem to have been both destroyed during World War II.

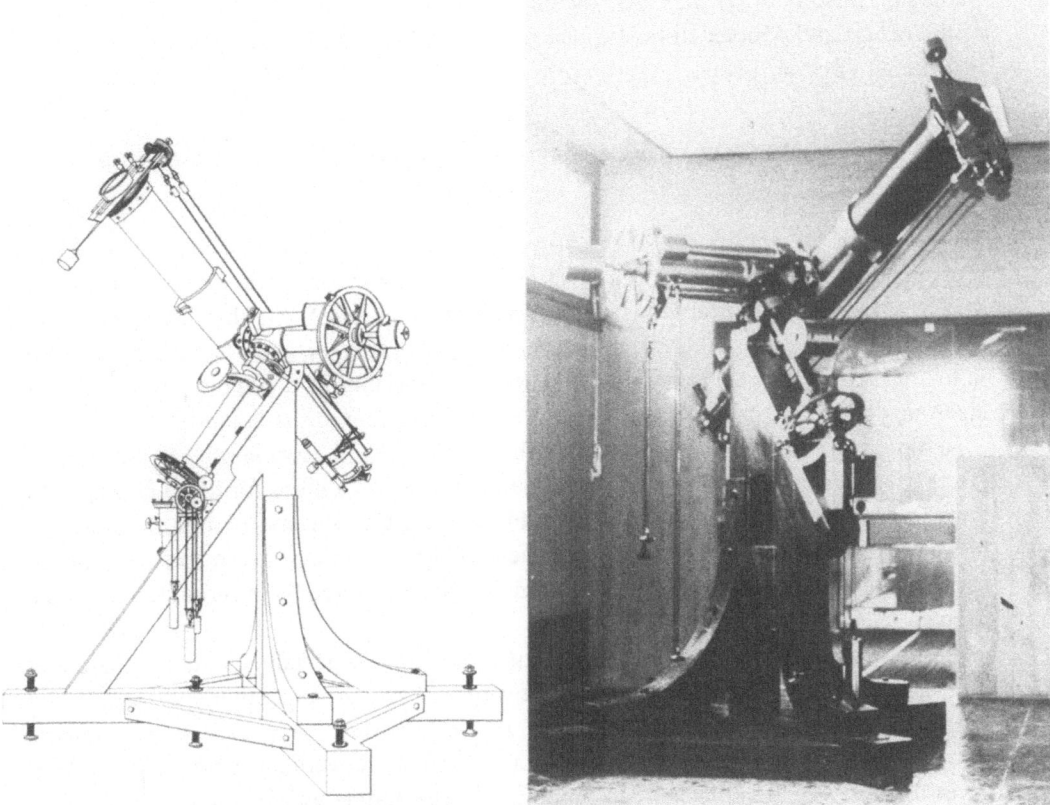

Fig. 5.   Bessel's large Fraunhofer heliometer (1829).

When the Königsberg instrument was finished in 1827–1929, Bessel travelled again to Munich to carry out first workshop tests. In a letter to Encke, director of the Berlin observatory who seems to have opposed Bessel on unknown reasons (jealousy?), he stressed that the instrument has been constructed strictly according to the plans of Fraunhofer. Furthermore, he urges him to have the instrument acquired for the Königsberg observatory, because 'no better one will be available elsewhere'.

The instrument finally arrived at Königsberg in 1829 and was immediately installed and tested on double stars, relative position measurements of Pleiades stars even in daytime, and the determination of the diameters of Saturn and of its rings. The accuracy (mean error) of a single observation was ± 0.15 arc sec for the distance and ± 2 arc min for the position angle. Under good seeing conditions he estimated the resolving power of the semi-lenses to be about 0.72 arc sec (Bessel, 1831).

On account of a series of measurements of the visual binary ρ Ophiuchi in 1837 he derived a mean error of the average of about ± 0.01 arc sec for the separation! In the same year Bessel (1838) carried out the famous observations for the parallax determination of 61 Cygni.

Finally, Bessel (1839) also used the heliometer for the determination of the libration of the Moon and concluded his heliometric investigations by a thorough theory of the instrument (Bessel, 1841).

3. Some Applications of the Heliometer Principle for Modern Observational Techniques

To the author the application of the bi-wedge method for image doubling seems to be most promising for increasing the positional accuracy by the outlined autocalibration of plates taken with shorter focus astrographic cameras and Schmidt-telescopes. Especially for the latter ones where in most cases the panoramic detector has to be squeezed onto the spherical focal surface, which is only possible in a limited range without stretching the detector's surface, this autocalibration method is essential.

In the few non-technical description of the HIPPARCOS astrometric satellite project (see, e.g., Kovalevsky, 1982) one misses autocalibrating facilities for focus control and to overcome the irregularities of the focal modulating grid system. Also in this case the bi-wedge offers its service in an arrangement schematically shown in Figure 6, as the entrance pupil of the Baker–Schmidt system is already divided by the complex astrolabe mirror.

Photometric autocalibration of panoramic detectors, especially of the photographic emulsion, by the bi-prism 'half-filter' arrangement has been recently developed by the author and collaborators (see, e.g., Geyer, 1984). The principle is to cover the cameras entrance pupil again with a bi-wedge one half of which is covered by a neutral absorbing layer. In this way the point spread functions of the double images are again identical, but the illumination of the one image is reduced by a constant factor. The characteristic

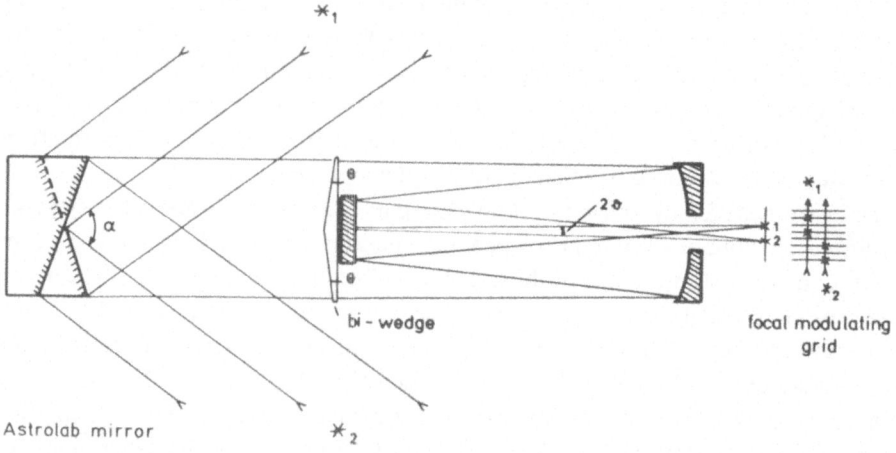

Fig. 6.   Proposal for the double-image autocalibration of the HIPPARCOS instrumentation package by a bi-wedge arrangement.

curve of the emulsion is then obtained from the iris readings of a series of such double images by the well-known Schwarzschild formalism.

Though already at the end of the 19th century the heliometer principle was also applied for astrospectroscopy, e.g., for the high-accuracy determination of the solar rotational velocity by projecting the opposite solar limbs with a heliometer onto the slit of a spectrograph, and a special double-image spectrometer with opposite dispersion direction for the relevant images was designed by Steinheil (see Ambronn, 1899), this method was not further persuited until recently.

At our institute we took up this idea, and designed a slitless field spectrograph, based on a focal reducer system and a double grating prism ('grism') in a roof-like arrangement, and which allows the measurement of absolute radial velocities by applying the Pickering reversion method simultaneously (Nelles, 1981; Geyer and Nelles, 1984). This can be performed according to the schematically layout of Figure 7: the identical halfs of the bi-grism are of the direct vision type for a certain wavelength λ_c ('coincidence wavelength'). The grating dispersion thus produces of a point source a pair of spectra with opposite direction of dispersion. If we consider the spectral lines as the monochromatic

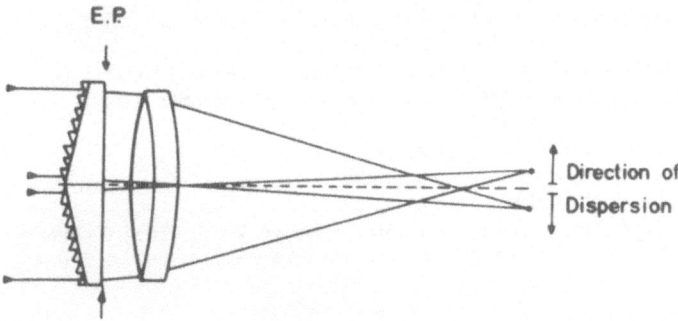

Fig. 7.   The bi-grism double image principle for absolute radial velocity determinations with a slitless field spectrograph.

images, then in the reverted spectra pair their relevant distances are constant for zero-velocity point sources. The variation of these distances are then due to twice the amount of the radial velocity of the object to which the optical distortion and dispersion errors over the field add. The latter can be calculated according to the grism attributes or be determined experimentically by a multislit mask at the telescopes focal plane illuminated by a laboratory spectral lamp (Nelles, 1984). It should be stressed that guiding errors and the seeing do not influence the accuracy, as was outlined previously. With such a focal reducer field spectrograph at a 1 m telescope was obtained absolute radial velocities of late B- to G-type stars with an accuracy of $\pm$ 6 km s^{-1} down to an apparent magnitude of 13^{m}5 using a reciprocal linear dispersion of 217 Å mm^{-1}. With a single exposure pairs of spectra of up to 40 stars can be obtained in the 25 arc min field of the telescope.

Acknowledgement

The support of parts of this overview by the Deutsche Forschungsgemeinschaft – grant Ge 209/10 – is gratefully acknowledged.

References

Ambronn, L.: 1899, *Handbuch Astron. Instrumentenkunde* **II**, 552.

Bessel, F. W.: 1812, *Monatl. Correspondenz* **XXV**, 285; **XXVI**, 148.

Bessel, F. W.: 1831, *Astron. Nachr.* **8**, 397.

Bessel, F. W.: 1837, in R. Engelmann (ed.), *Abhandlungen von F. W. Bessel*, Vol. II, Verlag W. Engelmann, Leipzig, p. 291.

Bessel, F. W.: 1838, *Astron. Nachr.* **16**, 65.

Bessel, F. W.: 1839, *Astron. Nachr.* **16**, 257.

Bessel, F. W.: 1841, in R. Engelmann (ed.), *Abhandlungen von F. W. Bessel*, Vol. II, Verlag W. Engelmann, Leipzig, p. 109.

Bouguer, P.: 1748, Bibliothèque impériale, Vol. III, p. 214.

Dollond, J.: 1753, *Phil. Trans. Roy. Soc. London* **48**, part I, p. 178.

Fanque, D.: 1983, *Rev. Histoire Sci.* **36**, 153.

Geyer, E. H.: 1984, *Roy. Obs. Edinburgh Occas. Reports* **14**, 279.

Geyer, E. H. and Nelles, B.: 1984, *Proc. IAU Colloq.*, No. 79 (in press).

Geyer, E. H. and Schmidt, H.: 1976, *Mitt. Astron. Ges.* **40**, 125.

Geyer, E. H., Hoffmann, M., and Nelles, B.: 1979, *Astron. Astrophys.* **80**, 248.

Kovalevsky, J.: 1982, 'Proceedings ESA-Conference of the Scientific Aspects of the Hipparcos Space Astrometry Mission', *ESAP-SP* **177**, 15.

Nelles, B.: 1981, *Mitt. Astron. Ges.* **54**, 186.

Nelles, B.: 1984, Doctor's thesis, University Bonn.

Nelles, B. and Geyer, E. H.: 1981, *Appl. Optics* **20**, 660.

Repsold, J. A.: 1908, *Zur Geschichte Astron. Messwerkzeuge*, Vol. I, Verlag W. Engelmann, Leipzig, p. 72.

Schur, W.: 1898, in W. Valentiner (ed.), *Handwörterbuch für Astronomie*, Vol. 2, p. 4.

BESSEL AND LIBRATIONS OF THE MOON*

KAROL KOZIEŁ

Jagiellonian University Observatory, Cracow, Poland

(Received 11 June, 1984)

Abstract. On the occasion of the twohundredth anniversary of F. W. Bessel's birth, his method of heliometric observations for the determination of the Moon's physical libration constants and of the reduction of these observations is presented.

Towards the end of the 18th century the founders of modern celestial mechanics, Lagrange (1780) and Laplace (1798), worked out the theory of the Moon's rotation and interpreted, on the basis of the law of universal gravitation, the empirical laws of Cassini concerning this rotation. These laws describe only approximately the rotation of the Moon. Thus, the irregularities of this rotation, treated as deviations from the laws of Cassini, are called the physical libration of the Moon. From the very beginning of the 19th century, attempts were made at finding from observations the constants characterizing the physical libration. This task was undertaken in 1806 by Bouvard and Arago on the basis of observations of the crater Manilius. Nicollet (1823) supplemented their observations and reduced the whole series of observations thus obtained, which yielded for the inclination of the Moon's equator to the ecliptic I the value $1°28'45''$, and for the Moon's mechanical ellipticity $f = B(C - B) : A(C - A)$, the value 0.055, where A, B, C, denote the lunar globe's principal moments of inertia. Soon afterwards Kreil (1837) and Stambucchi, who determined the position of crater Bode relatively to the Moon's limb, obtained $I = 1°35'48''$, and $f = 0.005$. The values of inclination I in the two enterprises do not, in principle, differ much from each other, but the quantities of the Moon's mechanical ellipticity f wholly remain out of any discussion and do not even constitute a first approximation. In the 1830's Beer and Mädler expressed the opinion that a crater as large as Manilius is unfit for investigation of the Moon's libration and, moreover, the said crater is lying rather far from the disk's centre.

One hundred and forty-five years ago the founder of modern astrometry, the Königsberg astronomer F. W. Bessel, worked out in full detail a method of observation to obtain the physical libration constants and published it in a paper entitled 'Ueber die Bestimmung der Libration des Mondes, durch Beobachtungen' (Bessel, 1839). He used for this purpose the heliometer, new at the time and improved by him, which has since become the standard instrument to be applied to the determination of the constants of the Moon's physical libration. In Bessel's opinion a profound knowledge of the Moon's libration is of great importance not only to selenography, but also for general-cognitive

* Communication presented at the International Conference on 'Astrometric Binaries', held on 13–15 June, 1984, at the Remeis-Sternwarte Bamberg, Germany, to commemorate the 200th anniversary of the birth of Friedrich Wilhelm Bessel (1784–1846).

purposes, all the more so, as it is connected with the problem of the Moon's figure and it may throw some light on the question of the solar systems origin.

In his method of observation Bessel choose the small crater Mösting A situated near the lunar disk's centre and very distinctly visible at various illuminations of the Moon, as the point on the disk, whose distances s from the illuminated lunar limb should be determined with the heliometer in various position angles p. In Bessel's paper, crater Mösting A is briefly called the point O. Thus Bessel advices to measure the distances of crater Mösting A in succession from seven points of the illuminated limb of the Moon's disc, uniformly distributed, reckoning from the limb's horns. Since each full measurement with the help of a heliometer requires measuring in both positions of the objectives halves, according to the procedure recommended by Bessel, the observer should pass from one horn of the Moon's limb to the other as uniformly as possible, i.e., to make the individual measurements in equal time intervals, and then to pass back again uniformly. Such manner of procedure requires, of course, a great observational skill, but owing to it, in the reduction of observations it can be accepted with great approximation that the measured distances s of crater Mösting A, i.e., of point O, as Bessel calls it, from the bright limb of the Moon's disk were observed simultaneously, which will extremely facilitate the subsequent reduction of observations of the given evening. Bessel also discussed the conditions, in which the said observations should be carried out, i.e., at what altitudes of the Moon above the horizon, in what kind of atmospheric conditions, and the like. Bessel also believed that his method of libration observations, on account of the great number of points of the limb, which he recommended to observe will, to a high degree, make the results, i.e., the constants of the Moon's physical libration, independent of the irregularities of the lunar limb. Unfortunately, this suggestion of Bessel verified itself only in part, and the later works, particularly since the days of Hayn (1914), started to use charts of isohypses of the Moon's limb, in order to correct the measured distances of point O from the limb for its irregularities, and thus to obtain correct values of the libration constants.

The method of libration observations outlined here after Bessel and published by him in 1839 was applied two years later in practice by his pupil Schlüter, who by the help of the Königsberg heliometer carried out a fine series of libration observations of the Moon covering a period of two and a half years and consisting of hundred-fifty eight evenings. After Schlüter's death in the last two years of Bessel's life, the heliometric observations were continued by his other pupil, M. Wichmann, in the years 1844 to 1846. In the second half of the 19th century, heliometric observations of the Moon with Bessel's method were undertaken by that indefatigable observer, E. Hartwig – the first director of the Dr Remeis Sternwarte Bamberg –, who observed the Moon for almost half a century and left us in his scientific heritage three series of observations. The first one is the Strassburg series, from the years 1877–1879, containing forty-two observation evenings, the second one, the Dorpat series, from the years 1884–1885, consisting of 36 evenings and the third one, the Bamberg series – the longest series ever carried out by one observer – consisting of 266 evenings over the period 1890–1922.

In his paper presenting the method of libration observations of the Moon, Bessel

(1938) states in the conclusion that, presumably, by the help of heliometers smaller than that in Königsberg, it will be possible to obtain equally valuable results as concerns the libration constant, because the observations with larger instruments are more liable to the disadvantageous influence of atmospheric air. The prophetic words of Bessel were splendidly confirmed by the observations conducted at the Engelhardt Observatory, where heliometric observations of the Moon for the determination of physical libration constants were carried out, with short interruptions, from 1895 to 1964 by the following astronomers: Krasnov (1895–1898), Michailovski (1898–1905), Banachiewicz (1910–1915), Yakovkin (1916–1931), Belkovich (1932–1942), and Nefediev (1936–1964). Over this time interval they collected the impressive number of over 1100 published observation evenings. And although the Kazan heliometer – constructed by Repsold in Hamburg primarily for observations of the transit of Venus across the Sun's disk in 1874 – is much smaller than the Königsberg heliometer, the results of the Kazan observations are by no means inferior to those of the Königsberg observations.

In the above-quoted paper published in *Astronomiche Nachrichten*, Bessel (1839) gives instructions concerning observations for the determination of the Moon's physical libration constants, and then presents the details of the reduction of these observations elaborated by himself on the basis of Lagrange's and Laplace's theory. In effect, each individual measurement of the distance s of crater Mösting A from the lunar disk's illuminated limb, in the set position angle p, gives one observation equation, the unknowns of which are the corrections to the libration constants, i.e., corrections to the coordinates of crater Mösting A: $d\lambda$, $d\beta$, dh, correction to the mean inclination of the Moon's equator to the ecliptic dI, correction to the Moon's mechanical ellipticity df, correction to the Moon's mean radius dR_0 and, possibly, the free-libration constants. In Bessel's days the derivation of such an observation equation, and particularly of the effective formulae for its differential coefficients, was somewhat complicated. Thus Bessel advised to solve this task in two stages: first to find the auxiliary unknowns of the problem – i.e., the corrections to the plane rectangular coordinates of crater Mösting A with the correction dR_0, and only the second adjustment yielded the proper unknowns of the problem, i.e., the corrections to the libration constants mentioned above. From the very beginning this manner of treatment afforded great difficulties in the choice of weights of the right-hand sides of the observation equations in the second adjustment. These right-hand sides are not independent, as they were obtained from the same equations in the first stage. But the adjustment of such equations in the second stage is not strictly correct from the point of view of the least-squares method. However, it should be pointed out here that the effect discussed was somewhat smoothed by the fact that Bessel used ecliptical coordinates in all this problem. In his paper published in the years 1948–1949, the writer (Kozieł, 1949) gave a method for adjustment of heliometric libration series comprising the observation equations of the problem in which, according to the suggestion of T. Banachiewicz, the so-called triangular cracovian square-roots were used. The correct and simple mathematical form of the final observation equations is given by the writer in 1956 (Kozieł, 1956, 1967). The left-hand sides of these equations contain the proper unknowns of the problem and the right-hand sides the independent quantities $(s_0 - s_c)$, s observatum minus minus s calculatum, which permits to carry out the adjustment in accordance with the principles of the least-squares method. The differential coefficients figuring in these equations can be found from our formulae, which were obtained on the basis of our differential formulae of spherical polygonometry (Kozieł, 1949).

Into the elaboration of the problem of determination of the physical libration constants, with which Bessel dealt in the last years of his life, he invested the experience of his activity in the field of astrometry abounding with excellent results. Still under his guidance, his pupil Schlüter was initiated in the heliometric observations of the Moon and soon afterwards Wichmann who, two years after Bessel's death, published in *Astronomische Nachrichten*, in Volume 27, the first results obtained with Bessel's method. Wichmann (1848) obtained for the mean inclination of the Moon's equator to the ecliptic the value $I = 1°32'9''$, practically identical – within the limits of their mean errors – with the value recently accepted $I = 1°32'4''$ (Kozieł, 1967; p. 23), and considerably different from Nicollet's value $I = 1°28'45''$ used in Bessel's days. Similarly, Wichmann's value of the fundamental libration constant $f = 0.48$ is comparable with the value presently accepted, and cancels Nicollet's undiscussible value $f = 0.055$.

In conclusion, I wish to point out that all our modern knowledge of the Moon's rotation about its centre of mass is predominantly based on observations with Bessel's method, whose principal assumptions have been applied by the observers for over hundred and twenty years.

References

Bessel, F. W.: 1839, *Astron. Nachr.* **16**, 257.
Hayn, F.: 1914, *Abh. Sächs. Ges. Wiss.* **33**, 1.
Kozieł, K.: 1948, *Acta Astron. Cracow* **4**, 61.
Kozieł, K.: 1949, *Bull. Acad. Polon. Sci.* **A**, 1.
Kozieł, K.: 1956, *Postepy Astron. Cracow* **4**, 78.
Kozieł, K.: 1967, *Icarus* **7**, 1.
Kreil, J.: 1837, *Effem. Astron. Milano.*
Lagrange, J. L.: 1780, *Mémoires de l'Acad. de Berlin.*
Laplace, P. S.: 1798, *Traité de Mécanique Céleste*, Vol. II, Paris.
Nicollet, M.: 1823, *Connaissance des Temps*, Paris.
Wichmann, M.: 1848, *Astron. Nachr.* **27**, 97.

THE HELIOMETER AS A TOOL FOR THE
MEASURE OF THE MOON*

MICHAEL MOUTSOULAS

University of Athens, Greece

(Received 13 June, 1984)

Abstract. The heliometer has been the only instrument for the measurement of the lunar physical libration for more than a century. Bessel (1839), who introduced the use of the heliometer for the systematic measurement of the relative positions of craters on the lunar disc, has also developed the necessary formulation for the calculation of the lunar physical libration from the heliometric measurements. That methodology is presented, and results obtained by Bessel's students and other investigators who followed Bessel's method, are discussed.

The heliometer was invented in 1748 by Bouguer, who thought that if an equatorial telescope could have, instead of one objective, two similar objectives mounted side by side, they would produce at the focal plane two images of the same star. If one of the lenses is made to slide on its plane, the corresponding image of the star will slide on the focal plane as well, and it can be arranged that the image of another star will fall on the former's position. Thus, the measurement of angular distances in the sky is converted into the measurement of the motion of the sliding lens. A few years after Bouguer's invention, Dollond modified the technique by replacing the two complete lenses by the two halves of one bisected lens, so that when the two semi-lenses are placed one against the other, forming a circular lens, only one image of a star appears at the principle focus, since the images that semi-lenses form are superimposed, while if the one semi-lens slides upon the other in a direction parallel to the line of section, the image it produces at the focus follows its motion, so that the distance between the two images formed on the focus is equal to the distance between the centres of the two semi-lenses. High-accuracy measurements of angular distances could be obtained with this double-image micrometer, and, since it was used for measurements of the diameter of the Sun, it was called 'heliometer'. Lalande (1771) used the heliometer for measurements of the diameter of the Moon.

Bessel (1839) introduced the use of the heliometer for the systematic measurement of the relative positions of craters on the lunar disc and, therefore, the detection of the lunar physical libration. Moreover, he shared Beer and Mädler's opinion that the main reason for inaccuracy in the observation of the lunar physical libration was the fact that large craters like Manilius, situated far from the centre of the lunar disc, were chosen

---

* Communication presented at the International Conference on 'Astrometric Binaries', held on 13–15 June, 1984, at the Remeis-Sternwarte Bamberg, Germany, to commemorate the 200th anniversary of the birth of Friedrich Wilhelm Bessel (1784–1846).

as reference points and he adopted their suggestion that a better point of the lunar surface should be chosen as reference for the measurements of the libration, and as such a point he proposed the crater Mösting A.

Let us consider on the celestial sphere two points, S and S', whose images have been brought into coincidence by appropriate movement of the heliometer's semi-lenses; if P is the celestial pole and S the middle of the arc SS', we find from the spherical triangles PS_0S and PS_0S' the equations

$$\sin \frac{s}{2} \sin p = -\cos \delta \sin (H_0 - H),$$

$$\sin \frac{s}{2} \cos p = -\sin \delta \cos \delta_0 + \cos \delta \sin \delta_0 \cos (H_0 - H),$$

$$\cos \frac{s}{2} = \sin \delta \sin \delta_0 - \cos \delta \cos \delta_0 \cos (H_0 - H);$$

and

$$\sin \frac{s}{2} \sin p = \cos \delta' \sin (H_0 - H'),$$

$$\sin \frac{s}{2} \cos p = \sin \delta' \cos \delta_0 - \cos \delta' \sin \delta_0 \cos (H_0 - H'),$$

$$\cos \frac{s}{2} = \sin \delta' \sin \delta_0 + \cos \delta' \cos \delta_0 \cos (H_0 - H');$$

where H, H', H_0, and δ, δ', δ_0 are the hour angles and declinations of S, S', and S_0, respectively, s is the angular distance SS' and p the position angle PS_0S'.

Moreover, if A is the pole of the heliometer's axis, and H_1, $.\delta_1$ are its hour angle and declination, while D is the arc AS and $q = P\hat{A}S$, we get from the triangle PAS the equations

$$\sin \delta = \sin \delta_1 \cos D + \cos \delta_1 \sin D \cos q,$$

$$\cos \delta \cos (H_1 - H) = \cos \delta_1 \cos D - \sin \delta_1 \cos D - \sin \delta_1 \sin D \cos q,$$

$$\cos \delta \sin (H_1 - H) = \sin D \sin q.$$

Similarly, the triangle PAS' gives

$$\sin \delta' = \sin \delta_1 \cos D' + \cos \delta_1 \sin D' \cos q',$$

$$\cos \delta' \cos (H_1 - H') = \sin \delta_1 \cos D' - \sin \delta_1 \sin D' \cos q',$$

$$\cos \delta' \sin (H_1 - H') = \sin D' \sin q'.$$

The above systems of equations give, finally

$$(r + r') \sin \frac{s}{2} \sin p = (u' - u) \cos(H_0 - H_1) - (v' - v) \sin \delta_1 \sin(H_0 - H_1),$$

$$(r + r') \sin \frac{s}{2} \cos p = (u' - u) \sin \delta_0 \sin(H_0 - H_1) +$$
$$+ (v' + v) [\cos \delta_1 \cos \delta_0 + \sin \delta_1 \sin \delta_0 \cos(H_0 - H_1)],$$

$$(r + r') \cos \frac{s}{2} = 2 [\sin \delta_1 \sin \delta_0 + \cos \delta_1 \cos \delta_0 \cos(H_0 - H_1)] -$$
$$- (u' + u) \cos \delta_0 \sin(H_0 - H_1) +$$
$$+ (v' + v) [\cos \delta_1 \sin \delta_0 - \sin \delta_1 \cos \delta_0 \cos(H_0 - H_1)],$$

where

$$u = \tan D \sin q, \qquad u' = \tan D' \sin q', \qquad v = \tan D \cos q,$$
$$v' = \tan D' \cos q', \qquad r = \sqrt{1 + uu + vv}, \qquad r' = \sqrt{1 + u'u' + v'v'}.$$

That system gives the angular distance and the position angle of the two points in terms of the measured coordinates.

In the case of lunar measurements, the position angles and the distances of points of the lunar limb are measured with respect to the reference crater. By comparing the so-observed value of the lunar libration to that, given by the theory, where some arbitrary value was given to the ratio of the mechanical ellipticities of the Moon α and β, Bessel proceeded to the correct value of f. The first heliometric observations of the lunar libration were carried out by Bessel's students Schlüter and Wichmann with the Königsberg heliometer. Reducing his observations, Wichmann (1846, 1847) was led to the value $f = 0.48 \pm 0.08$. Schlüter did not reduce his observations; this was done much later by Franz (1889) who produced the value $f = 0.488 \pm 0.028$, Stratton (1909) who obtained the value $f = 0.50 \pm 0.03$, and finally by Naumann who found $f = 0.71 \pm 0.08$ (Hayn, 1914). During the years 1875 and 1876, Hartwig, using the Strasbourg Repsold heliometer (focal length 170 cm, aperture 10.6 cm), carried out a series of 42 observations of the crater Manilius and in the next three years continued with another series of 42 observations of the crater Mösting A. Hartwig (1881) pointed out several errors which had been made previously in the reduction of the observations. A series of 36 observations was also made by him in Dorpat during the years 1884–1885 with a heliometer similar to the one he had used in Strasbourg, and between 1890 and 1922, 266 observations of Mösting A were obtained in Bamberg with the large Repsold heliometer (focal length 270 cm, aperture 18.4 cm). The first series of Mösting A observations was reduced first by Hartwig himself and led to the value $f = 0.507 \pm 0.060$ (Hartwig, 1881) and later by Franz (1887) who derived the value

$f = 0.4684 \pm 0.0564$. The series of Dorpat observations was reduced twice by Kozieł (1948, 1949), once with initial value $f_0 = 0.73$ and one with $f_0 = 0.5$. The results obtained were different lying on either side of the point 0.662. Naumann (1939) partly reduced the third series and found $f = 0.72 \pm 0.03$ without considering irregularities on the lunar limb, and $f = 0.71 \pm 0.03$ taking into account corrections for such irregularities.

The Kazan series of observations has played, also, an important role in the history of lunar measurements (Krasnov, 1895–1898; Michailovsky, 1898–1905; Banachiewicz, 1910–1915; Yakovkin, 1916–1931; Belkovich, 1932–1942; Nefediev, 1936 onwards). Völkel (1908) reduced Michailovsky's observations and found the value $f = 0.79 \pm 0.11$. Banachiewicz's observations were reduced by Yakovkin (1928), who found the value the value $f = 0.74 \pm 0.03$. Yakovkin is better known for his suggestion that the figure of the lunar disk does not correspond to that of a circle, but only the edge of the northern hemisphere is circular, while the edge of the southern hemisphere represents an ellipse of the form $R = R_0 + a \cos^2 p$ (Yakovkin, 1934). Yakovkin also reduced his own 251 observations of the lunar libration and found the value $f = 0.68 \pm 0.02$. In a new attempt (1950) to determine the constants of the Moon with allowance for changes in the lunar profile he found from the same observations $f = 0.82 \pm 0.03$. Belkovich reduced his 151 observations and found the value $f = 0.63 \pm 0.03$, while, repeating reductions of Michailovsky's observations he was led to the value $f = 0.84 \pm 0.08$. Belkovich, after considering the influence of the phase effect on his results, repeated the reduction of his own observations taking it into account, and obtained the value $f = 0.67 \pm 0.03$. A considerable number of observations has been carried out by Nefediev (1951, 1955, 1957, 1963) who, after reduction of his first 143 observations covering the period 1938–1945, found the value $f = 0.65 \pm 0.05$, using as initial value $f_0 = 0.73$ and $f = 0.57$ with initial value $f_0 = 0.60$.

The last, but by no means least, contribution to the detection of the lunar libration from heliometer observations is that of Kozieł (1948, 1949, 1962, 1964, 1967). Continuing Banachiewicz's work, he carried out the bulk of the reduction of 3282 observations of the Strasbourg, Dorpat, Bamberg, and Kazan series covering 370 evenings between 1877 and 1915. Having modified the reduction procedure used so far, Kozieł was led to the result $f = 0.633 \pm 0.011$.

In recent years much more precise methods for the measure of the Moon have been developed with use of photogrammetric techniques or space-age facilities (Moutsoulas, 1969, 1970). It should be recognized, however, that Bessel's heliometric method had been a significant milestone in the history of the measure of the Moon, a technique that dominated that field of research for more than a century.

References

Belkovich, I. V.: 1948, *Astron. Circ. USSR Acad. Sci.*, No. 81.
Belkovich, I. V.: 1949, *Izv. Engelhardt Obs. Kazan*, No. 24.
Bessel, F. W.: 1839, *Astron. Nachr.* **16**, 257.
Bouguer, P.: 1748, *Mém. Acad.*, p. 11.

Chauvenet, W.: 1891, *Manual of Spherical and Practical Astronomy*, Vol. II, Lippincott, Philadelphia, Pennsylvania.

De Lalande, J. A.: 1771, *Astronomie*, Vol. 2, p. 813.

Franz, J.: 1887, *Astron. Nachr.* **116**, 2761.

Franz, J.: 1889, *Astron. Beob. Königsberg* **38**.

Habibullin, S. T.: 1966, *Trans. Astron. Inst. Univ. Kazan*, No. 34.

Hartwig, E.: 1881, *Monthly Notices Roy. Astron. Soc.* **41**, 375.

Hayn, F.: 1914, *Abh. Saechs. Ges. (Akad.) Wiss.* **32**, 1.

Kopal, Z.: 1969, *The Moon,* D. Reidel Publ. Co., Dordrecht, Holland.

Kozieł, K.: 1948, *Acta Astron.* **4**, 61.

Kozieł, K.: 1949, *Acta Astron.* **4**, 153.

Kozieł, K.: 1962, in Z. Kopal (ed.), *Physics and Astronomy of the Moon*, Academic Press, New York, p. 27.

Kozieł, K.: 1964, *Trans. IAU* **12A**, 233.

Kozieł, K.: 1967, in Z. Kopal and C. L. Goudas (eds.), *Measure of the Moon,* D. Reidel Publ. Co., Dordrecht, Holland, p. 3.

Moutsoulas, M.: 1969, *New Scientist* **43**, 124.

Moutsoulas, M.: 1970, *Moon* **1**, 173.

Moutsoulas, M. D.: 1971, 'Librations of the Lunar Globe', *Physics and Astronomy of the Moon*, 2nd edition, Academic Press, New York, p. 29.

Naumann, H.: 1939, *Abh. Sächr. Akad. Wiss.* **43**, 1.

Nefediev, A. A.: 1951, *Izv. Engelhardt Obs. Kazan*, No. 26.

Nefediev, A. A.: 1955, *Izv. Engelhardt Obs. Kazan*, No. 29.

Nefediev, A. A.: 1963, *Bull. Engelhardt Obs. Kazan*, No. 30.

Stratton, F. J. M.: 1909, *Mem. Roy. Astron. Soc.* **59**, 257.

Völkel, M.: 1908, *Tr. Astron. Inst. Univ. Kazan*, No. 17.

Wichmann, M.: 1846, *Astron. Nachr.* **26**, 289.

Wichmann, M.: 1847, *Astron. Nachr.* **27**, 53, 81, 97, 211.

Yakovkin, A. A.: 1928, *Izv. Engelhardt Obs. Kazan*, No. 13.

Yakovkin, A. A.: 1934, *Bull. Astron. Inst. Engelhardt*, No. 4.

Yakovkin, A. A.: 1950, *Publ. Kiev Astron. Obs.*, No. 3.

CONCLUDING REMARKS

ZDENĚK KOPAL

Department of Astronomy, University of Manchester, England

In conclusion of the conference just past, may I be permitted to record a few thoughts which may also have occurred to many colleagues.

First, what would Bessel have thought if he could have been with us to hear how far the science, for which he laid the foundations in the first half of the 19th century, has progressed; if he heard what Drs Bernacca and Dommanget had to say on the impending Project HIPPARCOS, which promises to increase the accuracy of positional measures of the stars 100 times greater than that attainable by Bessel himself and his contemporaries; if he heard of laser ranging of the Moon, and of the studies of its librations with the aid of cube-corner retro-reflectors deposited by men on the surface of the Moon for this purpose?

He would, no doubt, have been astonished by such feats beyond imagination. And yet, do these and other advances accomplished in the time which elapsed since Bessel's time mean that we are so much brighter or more intelligent than our 19th-century ancestors? By no means! For any historian of science in recent times cannot fail to note that the principal reason why the advances accomplished since the time of Bessel – and, in particular, within our lifetime – have been due to no superior intelligence or understanding, but to the support of parallel advances in human *technology* at our disposal which made, of course, tremendous strides in the past 150 years. The motives of its magnificent efflorescence were not to advance basic science, but to apply existing knowledge to practical ends; yet is is undeniable that astronomers were important beneficiaries of this technological spin-off (and sometimes even prime contributors to it).

During Bessel's lifetime, observational astronomy was still essentially a visual art; with photography (terrestrial) making only its first steps when Bessel died in 1846; and electronics was still far in the future. The only technology which Bessel could avail himself of in support of his work was provided by mechanically-minded contemporaries like Josef Fraunhofer, Reichenbach-Ertel, Repsold, or Rheinfelder. Yet it is remarkable how far – with their aid – Bessel was able to *eliminate systematic errors* in his work, to obtain results astonishing in their accuracy even today.

The precision with which Bessel was able to determine the parallax of 61 Cygni with his heliometer was already paid tribute in my introductory remarks (p. 3). Another example of what his heliometer could do was provided, by his successors Schlüter and Wichmann at Königsberg (and especially Ernst Hartwig of Strassburg and Bamberg!) in their studies of physical librations of our Moon.

Of the parameters specified by the periods and amplitudes of lunar physical librations,

* Communication presented at the International Conference on 'Astrometric Binaries', held on 13–15 June, 1984, at the Remeis-Sternwarte Bamberg, Germany, to commemorate the 200th anniversary of the birth of Friedrich Wilhelm Bessel (1784–1846).

Astrophysics and Space Science **110** (1985) 203–207. 0004–640X/85.15
© 1985 *by D. Reidel Publishing Company.*

of paramount importance are the ratios

$$\alpha = \frac{C - B}{A} \,, \qquad \beta = \frac{C - A}{B} \,, \qquad \gamma = \frac{B - A}{C} \,, \tag{1}$$

where A, B, C denote moments of the Moon about its principal axes of inertia. The value of β can be determined, in principle, from the amplitudes of the physical librations (i.e., from the inclination of the Moon's orbit and equator to the ecliptic, and the rate of regression of the nodes of lunar orbit); and that of γ, from their observed periodicity; while the value of α corresponding to known β and γ can be evaluated from the identity

$$\alpha - \beta + \gamma = \alpha\beta\gamma \,. \tag{2}$$

The fact that the differences $C - A$, $C - B$, or $B - A$ (and, therefore, the ratios α, β, or γ) must be different from zero is amply attested by the observed synchronism between rotation and revolution of our satellite; for if the shape of the Moon were a sphere – with a spherically-symmetrical distribution of matter in its interior – the Earth's attraction would be powerless to effect synchronization. The existence of the latter cannot, to be sure, by itself specify the actual amounts by which α, β, or γ may differ from zero; this can come out only from a study of the librations performed by the distorted globe of the Moon about its centre of gravity. The arduous task of determining the values of β and γ from long series of heliometric observations (extending over almost one century) was carried out in 1962 at Manchester by Koziel (1967a, b), with the results disclosing that

$$\alpha = 0.000\,398 \pm 0.000\,008 \text{ (m.e.)} \,,$$

$$\beta = 0.000\,629 \pm 0.000\,001 \text{ (m.e.)} \,, \tag{3}$$

$$\gamma = 0.000\,231 \pm 0.000\,006 \text{ (m.e.)} \,.$$

It was noted many years ago (cf., for example, Jeffreys, 1924) that *the observed values of α, β, and γ are completely at variance with those to be expected if the lunar globe were in hydrostatic equilibrium* under the field of force to which the Moon could have been exposed. In particular, the actual value of β proves to be approximately 17 times as large as a hydrostatic one under the prevalent field of force; while α and γ are 42 and 8 times as large, respectively.

This is true, to be sure, at the present distance of the Moon from the Earth. This distance need not have remained constant throughout the long astronomical past of our satellite; and could once have been very much less – which would have influenced the absolute values of A, B, and C. But regardless of the possibility of bringing any one of the observed values of α, β, or γ in agreement with hydrostatic theory by allowing the Moon to have acquired its form in closer proximity of the Earth, the requisite proximity does *not* turn out to be the same for all three. In other words, the *ratios* of the observed lunar values of α, β, and γ prove to be inconsistent with the existence of hydrostatic (or lithostatic) equilibrium at *any* distance of the Moon from the Earth.

In order to demonstrate this, let us form the ratio

$$f \equiv \frac{\alpha}{\beta} = \frac{B}{A} \frac{C - B}{C - A} \; . \tag{4}$$

In hydrostatic equilibrium, this should be (cf., for example, Kopal, 1969) equal to

$$f = \frac{m_{\oplus} + m_{\mathrm{q}}}{4m_{\oplus} + m_{\mathrm{q}}} \; , \tag{5}$$

where $m_{\oplus}$ and m_{q} denote the masses of the Earth and the Moon, respectively; and this equation should hold good regardless of the Moon's distance or internal structure. As, moreover, the latest value of the ratio $m_{\oplus}/m_{\mathrm{q}}$ is equal (cf. Table 3–1 of Kopal, 1974) to 81.302 $\pm$ 0.001 (m.e.), the 'hydrostatic' value of f as given by Equation (4) should be equal to 0.2523; whereas the observed value of f consistent with Koziel's values of α and β is equal to

$$f = 0.633 \pm 0.006 \text{ (m.e.)} \; ; \tag{6}$$

and some previous investigators (e.g., Gorynia, 1965; Habibullin, 1966) have put it even higher.

These values are so much at variance with the requirements of hydrostatic equilibrium that the discrepancy must be regarded as real. Several thoughts should be considered in this connection. On the observational side, it is true that Koziel's results are based on heliometric measurements made with instruments of apertures smaller than 17.5 cm; and their Rayleigh limits of resolution in visible light was, therefore, not less than 0″.8. In order to obtain, from such measurements, results significant to 0″.1 (as quoted by heliometric observers), the mean of a great many individual settings must be made on the assumption that no systematic errors are present to impair their mean to this accuracy.

Positional astronomers have, to be sure, long been accustomed to search for the desired information inside optical diffraction patterns of their objectives; and have done so extensively when measuring, for example, stellar parallaxes. The success of such a process requires, however, a knowledge of the geometrical relation of the actual shape of the light source to that of its diffraction image (i.e., a point to a disc in the case of a star), in addition to a great many individual measures to minimize their accidental errors. In the case of selenodetic measurements the first condition cannot, unfortunately, be met; for the actual shape of lunar details on which heliometric settings are made are not known to us *a priori*; and neither is, therefore, the form of their diffraction image (which may, moreover, vary with the phase as a result of different illumination).

Under these conditions, many of us have been veritably holding our breath until the result of heliometric studies of lunar librations could be confirmed by independent methods. This has come to pass since the landing on the Moon of Apollo 11 in July 1969, when a cube-corner retro-reflector for laser signals was installed on the shores of the lunar Mare Tranquillitatis; followed by similar devices (of improved design) installed by Apollo 14 and 15, or the unmanned Lunas 17 and 21, in the proximity of their landing places. Observed echoes of laser pulses flashed on the Moon from the Earth and returned by these devices – permitting determinations of the instantaneous distance between the terrestrial transmitter and the particular cube-corner reflector on the Moon from the time-delays of the respective light-echoes with a precision of the order of one part in 10^8 – have opened a new epoch in several branches of lunar studies – including that of lunar librations.

At present these laser-tracking studies are still continuing; but the first published results (cf., Bender *et al.*, 1973; Williams *et al.*, 1973) led to the values of

$$\alpha = 0.000\,404\,3 \pm 0.000\,001\,1\,,$$

$$\beta = 0.000\,631\,1 \pm 0.000\,000\,4\,, \tag{7}$$

$$\gamma = 0.000\,226\,8 \pm 0.000\,001\,0\,,$$

virtually *identical* with those derived from ground-based studies of lunar librations by means of terrestrial telescopes; their mean errors are somewhat smaller, but of the same order of magnitude. The internal agreement between these two sets of the data obtained by completely different means (and subject to entirely different types of errors) is indeed excellent, and inspires full confidence in the correctness of the results. It vindicates brilliantly the patient efforts of investigators like Ernst Hartwig (1851–1923) who dedicated a major part of their lifetime to this exacting task; as well as the skill which inspired investigators like Koziel to extract the requisite information from long series of observations inherited from bygone days. The only other instance of a comparable feat in the history of science one can think of has been Chapman's success (cf., Chapman, 1918) in detecting the amplitudes of the lunar atmospheric tides from the barometric data measured in the equatorial belt of the Earth, whose individual errors were 10–100 times as large as the amplitudes sought after.

Moreover, one additional consideration may be adduced, which could explain at least partly why the achievements of Friedrich Wilhelm Bessel and of his contemporaries (among whom his friend Carl Friedrich Gauss holds an especially prominent place) were not followed up by their descendents with quite the same success. Consider Bessel as an example! His principal scientific accomplishments date to the last third of his 62-years long life – i.e., a time during which most of his successors of comparable prominence (certainly in the 20th century) would have been reduced to the role of administrators, or to public figures having little time or opportunity to pursue the promising work of their youth on which their early fame was based.

The 'brain-pollution' – largely self-inflicted – to which humanity became a victim in our times I had an opportunity to discuss on another recent occasion (Kopal, 1984); and a discussion of the causes and consequences of our self-inflicted stultification need to be repeated in this place. Instead, let me end with a few words concerning what the future may hold in store for us, and for our subject. That the onward march of human technology will not slow down – let alone come to a halt in our time – is obvious to any student of the subject at the present time – if anything, technology is the last manifestation of human activity likely to decline; and if so, where is astronomy likely to be in another two centuries – the time which separates us from the time when F. W. Bessel opened his young eyes in Minden on this old Universe of ours – now 200 years older?

With all refraction anomalies, or effects of thermal expansion of flexures of ground-based telescopes of Bessel's time now completely absent aboard the satellites operating in weightless conditions, how serious will be the effects of interplanetary or interstellar extinction on observations carried out beyond the confines of the terrestrial atmosphere; and how far shall we be able to see through interstellar haze in the extreme UV? How far will astronomy of double stars progress after observations of requisite precision will enable us to close the gap between 'wide' and 'close' binaries – which now increases with increasing distance from the Sun – and what all shall we yet learn about binaries with mass-ratios intermediate between those of stellar pairs and planetary systems? How many planetary systems exist in our neighbourhood, or in vaster domains of the Milky Way. In another 200 years we should know all this, and more – isn't it almost worthwhile to wait that long to see for ourselves?

References

Bender, P. L. *et al.* (12 co-authors): 1973, *Science* **182**, 229.

Chapman, S.: 1918, *Quart. J. Roy. Met. Soc.* **44**, 271.

Gorynia, A. A.: 1965, *Nauch. Dumka (Ser. Astron. Astrophys)*, Kiev.

Habibullin, Sh. T.: 1966, *Trudy Astron. Inst. Univ. Kazan*, No. 34.

Jeffreys, H.: 1924, *The Earth* (first ed.), Cambridge Univ. Press, Cambridge.

Kopal, Z.: 1969, *The Moon* (second ed.), D. Reidel Publ. Co., Dordrecht, Holland, p. 88.

Kopal, Z.: 1974, *The Moon in the Post-Apollo Era*, D. Reidel Publ. Co., Dordrecht, Holland, p. 79.

Kopal, Z.: 1984, in C. Wickramasinghe (ed.), *Fundamental Studies in the Future of Science*, Cardiff Univ. Press, p. 231.

Koziel, K.: 1967a, *Icarus* **7**, 1.

Koziel, K.: 1967b, in Z. Kopal and C. L. Goudas (eds.), *Measure of the Moon*, D. Reidel Publ. Co., Dordrecht, Holland, pp. 3–11.

Williams, J. G. *et al.* (4 authors): 1973, *Moon* **4**, 190.

LIST OF PARTICIPANTS

Bernacca, P.   Asiago Astrophysical Observatory, University of Padova, Asiago, Italy

Böhnhardt, H.   Remeis-Observatory, Bamberg, F.R.G.

Bues, I.   Remeis-Observatory, Bamberg, F.R.G.

Chapman, R. D.   NASA-Goddard Space Flight Center, Greenbelt, U.S.A.

Dommanget, J.   Observatoire Royal, Brussels, Belgium

Drechsel, H.   Remeis-Observatory, Bamberg, F.R.G.

Eichhorn, H. K.   Dept. of Astronomy, University of Florida, Gainesville, Florida, U.S.A.

Fracastoro, M. G.   Isituto di Astronomia, Universita di Torino, Torino, Italy

Fredrick, L. W.   Leander McCormick Observatory, University of Virginia, Charlottesville, Virginia, U.S.A.

Fricke, W.   Astronomisches Recheninstitut, Heidelberg, F.R.G.

Geyer, E. H.   Observatorium Hoher List, Universität Bonn, Bonn, F.R.G.

Gong, S. M.   Purple Mountain Observatory, Nanjing, China

Goyal, A. N.   Dept. of Mathematics, University of Rajasthan, Jaipur, India

Halbwachs, J. L.   Observatoire de Strasbourg, Strasbourg, France

Haug, K. H.   Remeis-Observatory, Bamberg, F.R.G.

Herczeg, T.   Astronomical Institute, University of Oklahoma, Norman, Oklahoma, U.S.A.

Ishida, G.   Tokyo Astronomical Observatory, Tokyo, Japan

Kimura, K.   Purple Mountain Observatory, Nanjing, China

Knigge, R.   Remeis-Observatory, Bamberg, F.R.G.

Kopal, Z.   Dept. of Astronomy, University of Manchester, Manchester, England

Liu, C.   Purple Mountain Observatory, Nanjing, China

Rahe, J.   Remeis-Observatory, Bamberg, F.R.G.

Rakos, K. D.   Institute for Astronomy, Vienna, Austria

Schiener, B.   Remeis-Observatory, Bamberg, F.R.G.

Söderhjelm, S.   Lund Observatory, Lund, Sweden

Soffel, M.   Lehrstuhl für Theoretische Astrophysik, Universität Tübingen, Tübingen, F.R.G.

Strohmeier, W.   Remeis-Observatory, Bamberg, F.R.G.

Strupat, W.   Remeis-Observatory, Bamberg, F.R.G.

Van Gent, R. H.   Sterrewacht Sonnenborgh, Utrecht, The Netherlands

Van Hamme   Dept. of Astronomy, University of Florida, Gainesville, Florida, U.S.A.

Van't Veer, F.   Institut d'Astrophysique, Paris, France

Wan, L.   Shanghai Observatory, Shanghai, China

Wood, F. B.   Dept. of Astronomy, University of Florida, Gainesville, Florida, U.S.A.

ANNOUNCEMENT

Astrometric Binaries

Editors: Zdeněk Kopal, Jürgen Rahe

Please note that a hardbound edition of this special issue of *Astrophysics & Space Science*, Vol. 110, No. 1 (March 1985), is available from the publishers.

ISBN: 90-277-1979-5   Prices: Dfl. 120,– / $39.50 / £30.50